Tourism, Heritage and Commodification of Non-human Animals: A Posthumanist Reflection

Tourism, Heritage and Commodification of Non-human Animals: A Posthumanist Reflection

Edited by

Álvaro López López

Institute of Geography, National Autonomous University of Mexico, Mexico

Gino Jafet Quintero Venegas

Institute of Social Research, National Autonomous University of Mexico, Mexico

Carol Kline

Hospitality & Tourism Management, Appalachian State University, USA

CABI

CABI is a trading name of CAB International

CABI
Nosworthy Way
Wallingford
Oxfordshire OX10 8DE
UK

Tel: +44 (0)1491 832111
E-mail: info@cabi.org
Website: www.cabi.org

CABI
200 Portland Street
Boston
MA 02114
USA

T: +1 (617)682-9015
E-mail: cabi-nao@cabi.org

A catalogue record for this book is available from the British Library, London, UK.

ISBN-13: 9781800623286 (hardback)
9781800623293 (ePDF)
9781800623309 (ePub)

DOI: 10.1079/9781800623309.0000

Commissioning Editor: Claire Parfitt
Editorial Assistant: Lauren Davies
Production Editor: Shankari Wilford

Typeset by Exeter Premedia Services Pvt Ltd, Chennai, India
Printed and bound in the UK by CPI Group (UK) Ltd, Croydon, CR0 4YY

Contents

Contributors vii

Acknowledgements xi

Part 1: Humanist Benefits and Animal Exploitation

1 **Introduction. Muted Torment: Speciesism, Tourism and Animal Commodification** 1
David A. Varela-Trejo

2 **Heritage and Commodification of Non-human Animals in Tourism: A Posthumanist Reflection** 7
Álvaro López López, Gino Jafet Quintero Venegas and Carol Kline

3 **The Legacy of Zoos? Moral Reasons against Considering Zoos as Heritage** 34
Gustavo Ortiz-Millán

4 **Fill Your Cup: A Multi-species Exploration of Balinese Kopi Luwak Agrotourism** 47
Jes Hooper

5 **Animal Trade and Dark Tourism in Sonora Market** 61
Jean Azcatl Pineda and Alicia Mariana Penélope Castro Pérez

6 ***Jallikattu*: Walking the Tightrope between Culture and Animal Welfare** 74
Javed Salim and Nusrat Yasmeen

7 **Bullfights, Tourism and Intangible Cultural Heritage in Mexico: A Dissonant Heritage** 85
Georgina Flores Mercado

8 **Bullfighting at the San Marcos National Fair: Animal Cruelty behind Tourism and Heritage** 99
Alejandro Morales Pérez and Brenda Martínez Velasco

Part 2: Navigating Animal–Human Relationships: Tension in Minimizing Damage

9 **Restoring Relationships of Respect with Eagle Relatives: Tribal-led Eagle Aviary Tours in Oklahoma Indian Country** 112
Bobbie Chew Bigby

10 A Posthumanist View on Theming the Axolotl for a *Pueblo Mágico* **128**
Gino Jafet Quintero Venegas, Álvaro López López and Carol Kline

11 Feline Saviours: Nekonomics, Stray Agency and the Kawaiification of Heritage Tourism in Japan **140**
Émilie Crossley and Johan R. Edelheim

12 The Heritage of the Cute: Commodifying Pandas in Urban and Rural China **156**
Yulei Guo

13 Contested Boundaries: Towards a Posthumanist Political Ecology Approach to Recognize Human–Deer Relationship as an Intangible Cultural Heritage **168**
Rie Usui and Carolin Funck

14 Conclusion. Shifting Animal Abuse in Heritage Tourism to the Paradigm of Building Multispecies Communities **185**
Leonardo Garavito-González, Jorge Iván Ruiz Barrera, Estephania Sepúlveda Perdomo, Tomás Arias Roncancio and Mateo Nicolás Rico Medina

Index **191**

Contributors

Tomás Arias Roncancio is a student of Philosophy at the Universidad Externado de Colombia. Email: tomas.arias@est.uexternado.edu.co

Jean Azcatl Pineda holds a degree in Geography from the National Autonomous University of Mexico, where he investigated the relationship between the exploitation of animals and their use as a cultural resource. He is currently pursuing a postgraduate degree in Geography, with a particular interest in critical heritage studies, critical animal studies and critical tourism studies as areas of research. In his research, he focuses on the animal issue in urban markets and gastronomic festivals as units of analysis. Email: jean.geohum@gmail.com

Alicia Mariana Penélope Castro Pérez earned her Master's degree in Geography from the Faculty of Philosophy and Letters at the National Autonomous University of Mexico. Currently, she is pursuing a PhD in Geography at the same institution. She also works as a high school teacher at the Modern American School in Mexico City. Her research interests are focused on the geography of tourism, dark tourism, critical heritage studies and spatial complementarity. Within these areas of work, she has developed case studies on Auschwitz-Birkenau in Poland and Elmina Castle in Ghana, among others. Email: hajiroi@comunidad.unam.mx

Bobbie Chew Bigby is an enrolled member of the Cherokee Nation and based in Tulsa, Oklahoma. Bobbie recently completed her PhD focused on comparative Indigenous tourism, culture and resurgence and is currently a Postdoctoral Fellow at the University of Waterloo (Canada). Her past research fellowships, including a Fulbright award and Rotary Peace Fellowship, have taken her to Indigenous Australia, Burma, Cambodia, China and India for research and community-based work. Bobbie is a co-editor/co-author of *Socialising Tourism: Rethinking Tourism for Social and Ecological Justice* (Routledge, 2022) and *A Local Turn in Tourism: Empowering Communities* (Channel View, 2022). Email: bbigby@uwaterloo.ca; bobbiechewbigby@gmail.com; bobbie.bigby@my.nd.edu.au

Émilie Crossley is a researcher at the Center for Advanced Tourism Studies (CATS) at Hokkaido University in Japan. She recently completed a JSPS Postdoctoral Fellowship for Research in Japan, which explored kawaii (cute) animals in tourism through multispecies ethnography with a particular focus on captive Ezo red foxes. Émilie holds a PhD from Cardiff University in Wales and has previously written about volunteer tourists' emotional responses to poverty, ecological grief in tourism and qualitative research methodologies. Email: emilie.crossley@elms.hokudai.ac.jp

Johan R. Edelheim worked for more than a decade globally in the hospitality and tourism industries before becoming an educator. Behind his research lies a deeply rooted aim to care for human and more-than-human others. His research focuses on (a) the Scholarship of Teaching and Learning (SoTL) in tourism Higher Education; (b) Critical Tourism Studies (CTS) connected to the climate emergency, identity, ethnicity and power relations; (c) foundational philosophies of tourism and hospitality, especially axiology; and (d) domesticated companion animal-related tourism. He teaches at the Graduate School of International Media, Communication, and Tourism Studies at Hokkaido University in Japan. Email: johan.edelheim@imc.hokudai.ac.jp

Georgina Flores Mercado is a researcher with the Institute for Social Research at the National Autonomous University of Mexico (UNAM). She teaches in the postgraduate programmes of Political and Social Sciences and Music in the same university. She has carried out research projects related to identity construction and musical traditions in indigenous and rural communities in Mexico. Her research interests include critical studies of heritage, heritage and cultural tourism in indigenous peoples, the construction of cultural identities through music and citizen participation in cultural policies. She has published books, book chapters and journal articles on these subjects, as well as coordinated recordings of traditional music and dances in Mexico. Email: gflores@sociales.unam.mx

Carolin Funck obtained her PhD from the Albert-Ludwigs University, Freiburg. She is Professor for Human Geography at Hiroshima University (Japan), Graduate School of Humanities and Social Sciences. Her research focuses on the development of tourism in Japan, sustainable island tourism and the rejuvenation of mature tourist destinations. She is the author of *Tourismus und Peripherie in Japan* and co-author of *Japanese Tourism*. Email: funckc@hiroshima-u.ac.jp

Leonardo Garavito-González has a PhD in Urban and Environmental Studies from El Colegio de México A. C., Master's in Population Studies and Professional in Government and International Relations from Universidad Externado de Colombia. He is a member of the research group 'Social Processes, Territories and Environment', and a Professor and researcher in social networks analysis, social movements and collective action, tourism and posthuman geographies and socio-environmental projects at the Human and Social Sciences Faculty of Universidad Externado de Colombia. Email: leonardo.garavito@uexternado.edu.co

Yulei Guo is an assistant researcher at the Chengdu Research Base of Giant Panda Breeding. Her research interest is human–animal relations in tourism. Email: yulei.guo@panda.org.cn

Jes Hooper is an Anthrozoology PhD candidate at the University of Exeter, a member of the IUCN Species Survival Commission Small Carnivore Specialist Group and the Founder of The Civet Project research initiative. Jes' research centres on human–civet relations in the Anthropocene, including the civet coffee production and tourism industries, civet companionship and conservation. Email: Jh1220@exeter.ac.uk

Carol Kline is a Professor and the Director of the Hospitality and Tourism Management program at Appalachian State University. Her research interests have historically focused on topics such as foodie segmentation, craft beverages, agritourism, tourism entrepreneurship and tourism in developing economies, but is now solely focused on animal ethics. She is founder of Fanimal Inc., an international non-governmental organization that helps individuals find animal-focused and environmentally-focused careers. Email: klinecs@appstate.edu

Álvaro López López has a PhD in Geography from the Arts Faculty of the National Autonomous University of Mexico (UNAM). He is currently a senior researcher at the Institute of Geography, UNAM, and also a Professor in the undergraduate and graduate programmes in Geography at UNAM, where he teaches Space and Tourism. His basic line of research is the geography of tourism, from which he has developed research projects about sex tourism, dark tourism and the intersection between the geography of tourism and the geography of animals, among others. He has been president of the Mexican Academy of Tourism Research. Email: lopuslopez@geografia.unam.mx

Brenda Martínez Velasco is a Master's student in Geography at the National Autonomous University of Mexico. Her research interests include the commodification of bullfighting in the context of tourism and the role of non-human animals in urban environments. Currently, Brenda is investigating the production of dolphinariums as places in Riviera Maya, Mexico. Email: mtzvcobrenda@gmail.com

Alejandro Morales Pérez is a Geography PhD candidate at the National Autonomous University of Mexico, where he has also worked as an Assistant Professor. He has taught courses on the geography of tourism. Alejandro's research focuses on a posthumanist perspective of bullfighting as a tourist attraction and an identity element of the San Marcos National Fair. Previously, he investigated hunting tourism and its territorial effects. Email: alejandro.mp@hotmail.co.jp

Gustavo Ortiz-Millán is research professor at the Institute of Philosophical Research, National Autonomous University of Mexico (UNAM), and member of the National System of Researchers of Mexico's National Council for Science and Technology. He holds a PhD in Philosophy from Columbia University in New York. He is the author of two books on bioethics and the editor of several books, the most recent of which are *Mind, Language, and Morality* (co-edited with J. Cruz, Routledge, 2018) and *Covid-19 and Bioethics* (co-edited with M. Medina, UNAM Press, 2021). He has published widely on topics of ethics, bioethics and, particularly, animal ethics. Email: gmom@filosoficas.unam.mx

Gino Jafet Quintero Venegas is an associate researcher at the Institute of Social Research, National Autonomous University of Mexico (UNAM), with a strong academic background in Geography. Additionally, he completed a postdoctoral fellowship at the UNAM's Program of Bioethics with the project 'Bioethics, Geography, and Zootourism: Ethical reflections on the use of non-human animals as tourist resources and cultural elements'. His research interests revolve around tourism and space, animal geography and animal ethics. Currently, he serves as a Professor of Animal Geography, Geography and Ethics, and Biocultural Heritage Management at the Arts Faculty, UNAM. Email: jafet@sociales.unam.mx

Mateo Nicolás Rico Medina is a sociologist from Universidad Externado de Colombia and a student of Geography at Universidad Externado de Colombia. He is a Master's degree candidate on Investigación en Educación con énfasis en Ciencias Sociales, a Hip Hop culture researcher and popular educator in Bogotá, Colombia. Email: mateo.rico@est.uexternado.edu.co

Jorge Iván Ruiz Barrera is a student of Management Business in Tourism and Hospitality at Universidad Externado de Colombia. He is a member of the research group in tourism 'Tyquy Uba' at the Management Business in Tourism and Hospitality Faculty. Email: jorge.ruiz03@est.uexternado.edu.co

Javed Salim is a Junior Research Fellow at the Department of Tourism and Hospitality Management, Jamia Millia Islamia (A Central University), in New Delhi, India. His area of research is heritage conservation and management and heritage tourism. Email: javed.jmi007@gmail.com

Estephania Sepúlveda Perdomo is a student of Anthropology at Universidad Externado de Colombia, an intern at Instituto Distrital de Patrimonio Cultural de Bogotá, Subdirección de Gestión Territorial del Patrimonio and a member of Editorial Committee of PAI Ethnographic Journal. Related interests include environmental, posthuman and multispecies studies. Email: estephania.sepulveda@est.uexternado.edu.co

Rie Usui is an Assistant Professor at the College of Sustainability and Tourism at Ritsumeikan Asia Pacific University in Oita, Japan. She has a multidisciplinary background in anthropology, primatology, ecology and tourism geographies. She obtained a PhD from Hiroshima University in

Japan in 2019. Her main research interest lies within the overlapping area of tourism studies and human–animal studies, with a focus on wildlife tourism in Japan. Her research spans an array of topics, including wildlife tourism management, the impact of tourism development on communities and human–animal relationships, and the experiences, motivations and behaviours of wildlife tourists. Email: rieusui@apu.ac.jp

David A. Varela-Trejo has a PhD in Anthropology from the National Autonomous University of Mexico. He currently works as a lecturer in the Bachelor's degree in Ethnohistory at the National School of Anthropology and History (ENAH). He is a member of the Research and Training Centre in Transdisciplinary Critical Studies (CIFECT). His fields of work include multispecies ethnography and theory, critical animal studies, antispeciesist and decolonial perspectives in Anthropology, critical studies of disability and studies of the Anthropocene. His current work is focused on researching the relationships between humans and other forms of life, and multispecies critical theory. Email: davidvrltrj@gmail.com

Nusrat Yasmeen is an Assistant Professor at the Department of Tourism and Hospitality Management, Jamia Millia Islamia (A Central University), in New Delhi, India. She is a lifetime member of the Indian Tourism Congress and Indian Hospitality Congress. Her research expertise is in the history of tourism and heritage tourism. She has 17 years of experience of teaching heritage and history, heritage tourism and hospitality. She also has coordinated faculty development programmes and international conferences, and has published works in the heritage domain. Email: nyasmeen@jmi.ac.in

Acknowledgements

This work was supported by UNAM-PAPIIT <IN303621>.

We express our gratitude to Luis Lorenzo Esparza Serra for his translation and style corrections made to several chapters.

Thank you to our reviewers, who spent time carefully reviewing the chapters in this volume. Their questions and suggestions were invaluable to making this a stronger body of work and contribution to the field of animal ethics: Giovanna Bertella, Hin Heemstra, Kris Hill, Claire Jean Kim, Dominic LaPonte, Johan Lindsjö, Rose Marie Menacho, Juan Carlos Monterrubio, Catherine Oliver, Mucha Mkono, Vasile Stanescu and Richard White.

And certainly, we offer our love and gratitude to the non-human animals we are fortunate enough to share our lives with: Chinicuil, Ursa, Booker and Dean.

1 Introduction. Muted Torment: Speciesism, Tourism and Animal Commodification

David A. Varela-Trejo*
Escuela Nacional de Antropología e Historia, Mexico City, Mexico

Abstract
This introductory essay offers a panoramic view of the use and involvement of animals in tourism. Through a review of every chapter, it establishes speciesism as both an analytic category in itself and a political positioning. The unifying idea of this book is the role of heritage as a powerful engine that reinforces the subaltern position of non-human animals and reifies their bodies for consumption.

Muted Torment: Speciesism, Tourism and Animal Commodification

Life is movement. From inception, our bodies move; movement governs our existence and makes it possible, from the cell upward to the most complex configuration at every level. Such configurations may or may not entail geographical long-distance shifts, but when they occur, movement becomes travel. Travel in Spanish (*viaje*) comes from the Latin *via* ('way'). In English, the etymology of 'travel' is no less interesting. According to the Merriam-Webster dictionary, the word derives from the Middle English *travailen/travelen*, which means both to toil or overcome and to travel. Considering both etymologies together, the image is one of exertion for those who open up roads by walking. Such an effort has never been exclusively or mostly human, for in their construction as much as in their operation, roads have always depended on the work and sacrifice of many non-human animals.

In what we still call 'modernity', with its techno-scientific, political and economic system, the world has shrunk. In a Western or westernized view, moving around the world in pursuit of joyful experiences of various kinds may seem natural, but the consequences for the planet and for animals in general are substantial. The practice of travelling on roads built for the purpose of consuming the 'otherness' is called tourism. In Spanish, *turismo* is a term borrowed from the English 'tour', which means spin or turn. The tourist is someone who goes around the world consuming and discarding, once and again.

The studies assembled in this book allow us to enter this modern 'world of tourist travels' from the particular posthumanist perspective of their authors, who bring to the fore the various non-human animals that participate in this specific form of global connectivity.

Throughout the chapters of this book, the authors invite us to explore the assemblages – modes of relations among beings (Tsing, 2021a) – in which animals are objects for

*davidvrltrj@gmail.com

 Tourism, Heritage and Commodification of Non-human Animals (eds A. López López *et al.*)
DOI: 10.1079/9781800623309.0001

consumption during a tourist experience. The 'animals of tourism' are themed – that is, made conspicuous – by capitalist relations for the production of culture, exoticism, cruelty and tenderness. These texts also open a window to see the world from other perspectives, from which we can learn how to continue living on this damaged planet. The 'contributors' and co-authors of this book are bulls, palm civets, royal eagles, axolotls, panda bears and cats. This is a collective effort to build meeting points, to reflect and adopt a posthumanist practice that rejects the reification, commodification and exploitation of animals' bodies by tourism. It is about the complex and contradictory realities in which the capitalist logic, in combination with ontological, religious, identity, economic, ethical, political, historical and socio-cultural elements, is very difficult to offset. Though the historical processes of dominance are not the same everywhere, and do not have the same material effects, all the chapters converge on a direct or indirect analysis of heritage – the arbitrary selection of historical–cultural elements used to generate processes of production of 'animality' in dispersed tourism geographies and to stage situated domination performances – and the interest in finding alternatives to the management of nature and animals by institutions and the tourism market.

The book takes us on a journey to different tourist destinations that capitalize on the idea of heritage, starting at the zoo (Chapter 3), where, in addition to the presentation of historical arguments about the legitimacy of this institution, an assessment of its scientific–social premises is conducted. For example, zoos' preference for mammals and how charisma plays an important role in the selection of species that must be protected, whether or not they are endangered. Rather than taking animals to the zoo, a better alternative would be to protect their habitats and guarantee the integrity of their territory. In Mexico City, we visit Sonora Market (Chapter 5), where the overcrowding of live animals, their slaughter for ritual and religious purposes, and the unstated morbidity make for a dark and dissonant form of tourism, whose main attraction is the death of the 'other', the consumption of dead bodies as both image and meat to enjoy the exoticism of a truly 'Mexican' experience. As far as tourism is concerned, animals matter only inasmuch as they serve a utilitarian, ritual, culinary or entertainment purpose, and distance must be kept between them and the tourist.

The next stop in our imaginary journey is the so-called *fiesta brava* or bullfight, which despite their obvious decline are still celebrated in the Mexican states of Colima, Hidalgo and Aguascalientes, the latter with its famous San Marcos National Fair. The ideology of bullfighting pushes hard for the continuation of this controversial form of 'art', and under the aegis of anthropocentric institutions such as UNESCO and the economic support of the elites, its narrative manages to legitimize this form of torture as 'culture' and institutionalize it for posterity. As a dark and perverse form of tourism in which the staging of a supposed 'fight' creates a cognitive distancing between the bull and the spectator, bullfighting has earned a generalized social rejection that is increasingly harder to hide or silence by its supporters and their strategies (Chapters 6 and 7). On the other side of the world, in Tamil Nadu, India, another animal–human encounter is staged in celebration of nature's bounties, in which men must tame a bull and remove a bundle of coins attached to its horns as their prize (Chapter 5). Strongly anchored in religious feelings, in cultural identity and notions of male virility, this tradition is now challenged by NGOs and other forces, including the judiciary, which has created a friction scenario around cultural rights and animal rights.

The common thread that runs through the chapters of this book is the invention of tradition for the benefit of tourism. The arbitrary selection and organization of cultural traits lead to the production of stereotypes over human and non-human elements that are presented as natural. On the island of Bali, Indonesia, cultural authentication ('purification') processes take place in a multi-species framework (Chapter 4). Palm civets, coffee seeds, faeces, infrastructure, cages and humans participate in the staging of a sophisticated agrotourism experience: the production of kopi luwak, the famous and expensive coffee obtained through the digestion and defecation of coffee cherries by palm civets. However, due to the fact that the animals in question are caged and that the collection of the extremely valuable faeces is done in confinement rather than free range, the authenticity of the cultural experience tumbles and the tourist feels duped and cheated.

The controversy on what is 'cultural' and what 'natural', and whether an animal belongs to one or the other categories, and if it is ethically acceptable to commodify animals, has become even greater as human exceptionalism and the primacy of the global North over other societies began to be questioned. The countries of the global South have asserted their right to participate in the economic benefits of tourism and the progress it promises on their own terms. The widening of the horizons for subaltern groups brought about by modernity situated the lives of non-human animals in a relevant political and economic scenario. The artificial and arbitrary borders between nature and culture, human and animal, revealed themselves as more mobile than they were previously regarded by traditional thinking. For the humans and animals that had been historically subordinated to colonial rule, the question of who decides what tradition is – under what conditions and how this affects non-human animals – became central to the design of a tourist policy that privileges seeing (i.e. consuming visually) over the other sensorial faculties in providing cultural human–animal contact experiences.

This is the case in the USA, where royal eagles demand from us a new way to relate to life in general, based on respect. The task concerns everyone, and not only those who are directly responsible for the damage inflicted. The Citizen Potawatomi Nation in Oklahoma, USA, is developing 'resurgence paradigms' that emphasize the importance of Indigenous peoples re-establishing and strengthening their traditional connections to other forms of life, including the lands and water, and the community in itself, above any other interest (Chapter 9). The eagles fulfil the function of legitimizing a culture based on ancestral stories associated with them. Thus, they participate in a cultural identity process and the Citizen Potawatomi people can offer alternative tourism models that motivate eagle care through cultural education. In another geographical and cultural location, Chignahuapan, in the Mexican state of Puebla, the axolotl is a tourist product resulting from the reification of the amphibian to suit the 'tertiarization' of the economy by the abandonment of agricultural activity. The axolotl's presence is used as an identity element to theme the 'magical town' of Chignahuapan, that is, to stage a fiction through – once again – the selection of narratives and cultural features that prove competitive in relation to other towns so denominated. The body and charisma of the axolotl are used to offer the tourist a reality tailored to their fantasies of what it is to be Mexican. The axolotl undergoes a teleological transformation, that is, its life is caught in the ethical contradiction between collecting money from visitors and ensuring the preservation of the species through the promotion of alternative ways of relating to humans (Chapter 10).

In both these cases, there is a relative discourse that orbits around tourism and has a magical effect on tourists, who are taken to believe that their presence is something good, that it is thanks to them that axolotls, royal eagles or even palm civets and fighting bulls continue to exist, a modern ratification of colonial power. Conservation seems to obliterate the subjection of these animals' bodies. This is another way of purifying the tourism activities of the Anthropocene or Capitalocene, with their destructive force on other species and on the entire planet. This is an era defined by the dominion of the human 'species' over the planet – though not without dissent, as shown by the strong arguments of Ailton Krenak, reporter, activist and promoter of the rights of the indigenous peoples of the Amazon, who poses a provocative question: 'Are we really a humanity?' He wonders whether many of the beings that exist in this world are simply considered as a mass without their individualities being valued (Krenak, 2021).

The dominant civilizational and economic model has pushed humanity to a dead end in which nobody can be safe anymore. Though the implications of this cannot even begin to be sketched here, the one thing we can be clear about is that the protection and conservation of animal species is no longer limited to the so-called wild animals, such as giant pandas. In fact, the presence of 'domestic' animals such as cats can reactivate places apparently condemned to oblivion and ruin. And they actually do it (intentionally or otherwise) through the liberation in humans of endorphins and dopamine that move us and make us feel emotionally healthy.

We now travel to Kishi station, on the Kishigawa railroad line in Wakayama Prefecture, Japan, where tourism has been reactivated thanks to the commercial exploitation of Tama, a female calico cat that was adopted by the station managers. Tama's presence, until her death in 2015, was used to make the station and the whole railroad line an enchanting tourist destination. Today, the job of keeping alive the tradition and added value of nekonomy relies on Tama's successors, Nitama and Yotama, who unlike her, cannot be touched and work in cages 6 hours a day (Chapter 11). Tenderness is a social construct that has saved the giant panda of China from extinction. Their charisma is such that the likeness of the panda has been adopted as the logo of an important nature conservation non-profit organization, the World Wildlife Fund. Meanwhile, the difficulties experienced in panda reproduction (in line with the aura of extinction that has characterized them for decades) have turned pandas into a living cultural and natural heritage element whose bodies have been commodified for the promotion of rural and urban tourism in China. Cuteness-based heritage functions through complex emotional operations that activate affective economies, so the 'cutified' pandas are turned into consumption nodes sustained by the uniqueness of their biology (including their reproductive difficulties) and the discourses that surround it (Chapter 12).

In both cat and panda cases, the question arises as to whether the social construction of cuteness makes conservation more effective and how this affects tourists' subjectivities, leading them to consume according to their feelings, which have been modelled by tourism and heritage. So far, tourism seems to try to strike a balance between the theming and representation of animals and the sovereign gaze of the tourist. Conflicts between tourists and animals can occur when the latter misbehave and create interference in a tourist experience. Similar to cuteness, wildness is also a social construct.

The study of tourist–animal relations throws new light on some of the presuppositions on the nature of wildlife. If we travel to the Itsukushima Sanctuary on the island of Miyajima, Japan, with its red *torii* gate, we notice that the landscape cannot be understood without the deer. The deer participate in both landscape and social multi-species relations. Far from being pacific landscapes, though, conflict emerges the moment deer fail to behave according to the administrative blueprint of the tourist enclave (Chapter 13); not that tourism is intent on presenting animals as neutral species either. In fact, deer, palm civets, bulls, the axolotl, cats and any other tourism-themed animal are culturally and biologically constructed as wild or domestic. A deer is regarded as part of heritage or as an intruder, depending on the structurally unequal relations they have historically maintained with humans, as happens with the other animals that call our attention in this book.

Space is running out and the trip must end – at least for now. But there is still an important question to ask: what makes all this possible? Why is it that we name and describe the insidious forces that run through lives, bodies and subjectivities, turning some animals into 'tourism animals' (Kline, 2023) and others as animals for meat or laboratory experimentation? Though posthumanism – with its undeniable epistemological value – turns the world inside out to show the folds of modernity, the truth is that behind each page there is a political positioning. One of the important findings of posthumanism and the feminist epistemologies that adhere to it is precisely that scientific neutrality is an illusion used by those who exercise a form of knowledge to reify the other and speak in their name through the so-modern notion of scientific evidence as true testimony. Knowledge-producing bodies are always situated, that is, they respond to their own conditions of existence, and thus all scientific facts are always made, fabricated, modelled (Haraway, 2019). We then lack a word to name the anthropocentric machinery that dictates a hierarchical functionality and normativity over every living being in the world (Ávila, 2012).

The hierarchical structuring of the earth's creatures responds, in the first place, to human power, which any individual can exercise by the simple fact of belonging to the human category, and which is the product of correlations within an oppressive social order (González and Ávila, 2022). This specific mode of oppression that bodies classified as 'animal' suffer is called 'speciesism'. In its classical or traditional sense, speciesism is understood as a form of discrimination. However, it is far more useful for this book to understand it as a global biophysical order that systematically produces and reproduces animal domination, exploitation and subjugation through the exercise of human power (Ávila, 2012). In this sense, this volume's contributions contain plenty of examples of the functioning and reproduction of this order; subjects, objects, practices, forms of knowledge and relations that work through devices such as zoos, vivaria and industrial farms (González and Ávila, 2022). In its practical relations, speciesism is not only discriminatory but effectively turns animals into workers, objects, heritage pieces and practices embedded in contexts that, though 'fictitious', have nevertheless real effects on the animal bodies it seizes. Tourism makes it possible to mobilize and actualize the speciesist order through the satisfaction of tourists' desires using animal bodies.

In their specific forms, the 'tourism animals' (palm civet, bull, deer, eagle, axolotl, cat) embody cultural functions for the satisfaction of tourists' expectations, informed as they are by the capitalist-speciesist order that turns their lives into a resource for consumption under the protection of a modern colonial-speciesist heritage. The texts contained in this book are a provocation to think of animal realities, not to talk in their stead but to denounce speciesism. This is an outstanding contribution to an academic field in which animals tend to appear only as support of human activity and in which their presence is important only insofar as it helps understand some exclusively human activity.

Taking these texts as an example, I postulate that speciesism functions as a constant reality in friction that makes it possible to get hold of cultural diversity. Speciesism, in other words, is in constant evolution alongside the global and local aspirations related to entertainment, enjoyment and exotic animal experiences that tourists seek to satisfy at their destination. As a result, new cultural formations are created that allow speciesism to integrate into various social contexts, establishing bridges, paths and channels for the circulation of non-human and human lives (Tsing, 2021b).

The way tourism utilizes animals and the specific forms it takes are closely related to natural, cultural and heritage diversity, and the interactions between multiple species that arise during encounters between tourists and locals. In this way, animal exploitation is universal, strongly accentuated by modernity. As global order, speciesism is part of global connectivity that informs us about the materiality of animal–tourist encounters and how this modifies their trajectories.

A counter- and anti-speciesist perspective is then an explicit political positioning that admits that the various forms of life we call animals and with whom we exist do not have the same preferences as our own. Celebration of difference indeed – and of resistance! Though not all the authors of this book necessarily share this position, their contributions advance a strong insight towards a future of multi-species justice for all.

In brief, this volume invites us to ask ourselves whether tourism, independently of the scenario of production of 'the animal', must undergo reform. That is, to practise a welfare-oriented policy concerned with the interests of animals and revisit forms of situated speciesism. The aim being to make them more akin or in tune to certain sensibilities, even if the logic of animal exploitation and economic interests remain untouched – or to promote situated forms of abolition of animal subjection, exploitation and domination at all levels.

We could move against speciesism actively or leave it to the fullness of time to define how we respond. In any case, the inevitability of contact and multi-species relations and future responsibilities are two among many other possible important points that challenge us to ask ourselves about the possibilities of continuing to inhabit this earth. In short, how are we to reorganize life from the damaged ways that tourism has handed down to us? Or, paraphrasing anthropologist Lévi-Strauss (2008), what would the 'common heritage of humanity' be if we recognized that our humanity is defined by our relationship with other humans, non-humans and all other forms of inanimate materiality that form part of nature?

References

Ávila, I. (2012) *De la Isla del Doctor Moreau al Planeta de los Simios: La Dicotomía Humano Animal como Problema Político (From Doctor Moreau's Island to the Planet of the Apes: The Dichotomy Human–Animal as a Political Problem)*. Ediciones desdeabajo, Bogotá.

González, A. and Ávila, I. (2022) *Glosario de Resistencia Animal(ista) (Glossary of Animal[ist] Resistance)*. Ediciones desdeabajo, Bogotá.

Haraway, D. (2019) *Las Promesas de los Monstruos. Ensayos Sobre Ciencia, Naturaleza y Otros Inadaptables (The Promises of the Monsters. Essays on Science, Nature and Other Inadaptables)*. Holobionte Ediciones, Salamanca, Spain.

Kline, C. (2023) Centering animals within rural tourism. In: Mair, H. (ed.) *Handbook on Tourism and Rural Community Development*. Edward Elgar Publishing, Cheltenham, UK, pp. 344–358.

Krenak, A. (2021) *Ideas para Postergar el Fin del Mundo*. Prometeo Libros, Buenos Aires.

Lévi-Strauss, C. (2008) El Occidente puede aprender mucho de Asia (the West can learn much from Asia). *El Correo de La UNESCO* 5, 12–16.

Tsing, A. (2021a) *Fricción. Una Etnografía de la Conectividad Global (Friction. An Ethnography of Global Connectivity)*. I.F. Publications, Barcelona, Spain.

Tsing, A. (2021b) *La Seta del Fin del Mundo. Sobre la Posibilidad de Vida en las Ruinas Capitalistas. (The Mushroom at the End of the World. On the Possibility of Life on the Ruins of Capitalism)*. Capitán Swing, Madrid.

2 Heritage and Commodification of Non-human Animals in Tourism: A Posthumanist Reflection

Álvaro López López[1]*, Gino Jafet Quintero Venegas[1] and Carol Kline[2]
[1]Universidad Nacional Autónoma de México, Mexico City, Mexico; [2]Appalachian State University, Boone, North Carolina, USA

Abstract
This chapter, based on an analysis of texts, reflects on the involvement of non-human animals in tourism from a post-anthropocentric and anti-speciesist posthumanist philosophical perspective. It explores two aspects: when animals are included in heritage designation (natural, mixed or cultural) and when they are involved in heritage environments. The research aims to highlight and deconstruct philosophical discourses that normalize and sustain the commodification of animals, disregarding their fundamental interests. Furthermore, based on the UNESCO heritage lists, some examples are shown where non-human animals are objectified in tourism.

Introduction

Homo sapiens is an animal species like any other. The use of the qualifier 'human' in this chapter to refer to the members of this species does not imply in any way or measure that we consider them to be less animal than others, and responds merely to practical purposes. Our study is a theoretical exercise on the ethical issues surrounding the designation of animals in general as part of the cultural 'heritage' of tourist destinations from the perspective of post-anthropocentric and anti-speciesist (Faria, 2016) philosophical posthumanism (Ferrando, 2019). It proceeds through the analysis of texts about animal categories, heritage and tourism that contain in some measure a posthumanist and anti-speciesist philosophical insight.

The chapter is organized in four sections. The first section reflects on the differences between the moral consideration of animals in humanism on the one hand and philosophical post-anthropocentric anti-speciesist posthumanism on the other, evidencing the power relations of humans towards animals. The second section presents heritage as a cultural construction that normalizes and perpetuates patriarchal power relations, which allows for animal exploitation. The third section analyses the ethical approaches that have provided the foundation to multiple studies on the animals–tourism interaction and the way in which animals have been reified by the tourism system. Finally, the fourth section discusses the ways in which heritage designation by UNESCO (2023a, b) has contributed to the reification and commodification of the lives and bodies of animals in tourism.

*Corresponding author: lopuslopez@geografia.unam.mx; lopuslopez@yahoo.com.mx

DOI: 10.1079/9781800623309.0002

From Humanism to Posthumanism: Changes in the Moral Status of Animals

Posthumanism is an umbrella term bringing together a number of philosophical insights on the meaning of being human amid humanism's crisis (Chavarría, 2015). Among the most popular of these insights are: antihumanism, which deconstructs the notion of the individual and questions their autonomy (Ferrando, 2013); transhumanism, which addresses the techno-scientific knowledge of the human body (Bostrom, 2005; Weaver, 2009); cosmism, which advocates the replacement of humans by artificial intelligence (De Garis, 2005); voluntary human extinction, which presents itself as the best option to attain a true ecological balance (Ormrod, 2011); and philosophical posthumanism, which seeks to escape human exceptionalism (implied in humanism) or the idea that humans enjoy special privileges over other animate beings (Ferrando, 2013, 2019). What follows is a presentation of the basic tenets of philosophical posthumanism.

Philosophical posthumanism constitutes a departure from humanism, which advocates the centrality of human beings situated in a position of superiority *vis-à-vis* their biological others (Wolfe, 2010). This notion has been made possible by the establishment of power devices such as the nature–humanity dichotomy. Humanism's more sublime achievements, culture and civilization are founded on the belief that human beings can be abstracted from 'nature' and divested of their animality, so that they can control their own instincts and transcend barbarism (Elias, 2000; Wolfe, 2010; Quintero and López, 2020; López and Quintero, 2021a).

The negative appraisal of animality was strongly reinforced by René Descartes' ideas on animals (Cottingham, 1978), whose influence runs deep in Western thought and has ever since dominated the way in which animals are generally perceived within human society. The Cartesian argument is that animals are like machines devoid of consciousness and the ability to experience pain or emotions. Human–animal relationships have been shaped by these notions and, in a rather utilitarian and unethical fashion, humanism has legitimized the use of animals as resources with no regard for their rights. The contemporary practice of making animals part of the heritage (Sp. *patrimonio*) of local cultures turned into tourist destinations has but normalized the disregard for their intrinsic value as individuals (Miller, 2013).

According to Ferrando (2019), within philosophical posthumanism are lodged efforts to deconstruct humanism's Eurocentric, rationalist, anthropocentric, dualist and patriarchal assumptions (Castree and Nash, 2006; Wolfe, 2010; Faria, 2016; Urbanik and Johnston, 2017; Cudworth *et al.*, 2018; Ferrando, 2019), and though their common purpose is to decentre human beings as the exclusive owners of moral consideration, not all of the existing philosophical positions (ecocentrism, biocentrism and zoocentrism – see Fig. 2.1) have attained a congruent and conclusive argumentation. This chapter will address the discussion on the ethical viability of animal commodification by the tourism industry from the point of view of zoocentrism.

While ecocentrism extends moral consideration to biotic and abiotic communities such as plants, fungi, bacteria, soils, water and air, among others (Taylor, 2011; Cancino and De Alba, 2018), and biocentrism considers communities of living beings and other organic components of ecosystems (Kaltofen, 2018; Cole-Turner, 2022), zoocentrism focuses on the intrinsic moral value of humans and other animals as individuals (Singer, 1975; Francione, 2000; Horta, 2007; Faria and Paez, 2014). The philosophical positioning of ecocentrism, biocentrism and zoocentrism derive, respectively, into ecoethics, bioethics and zooethics.

In the moral equality among communities postulated by ecoethics and bioethics, the human species enjoys, more often than not, a position of exceptionality (Sass, 2007; Muzur and Rinčić, 2015; Faria, 2016; Cancino and De Alba, 2018). Thus, when the notion arises that the unusual growth of an animal or plant species has created an imbalance, its annihilation is seriously considered. Such a 'solution' would hardly ever be suggested for the human species, despite the fact that its unhampered growth and endless consumption have had and continue to have a manifest impact (both directly and indirectly) on ecosystems and the equilibrium among species

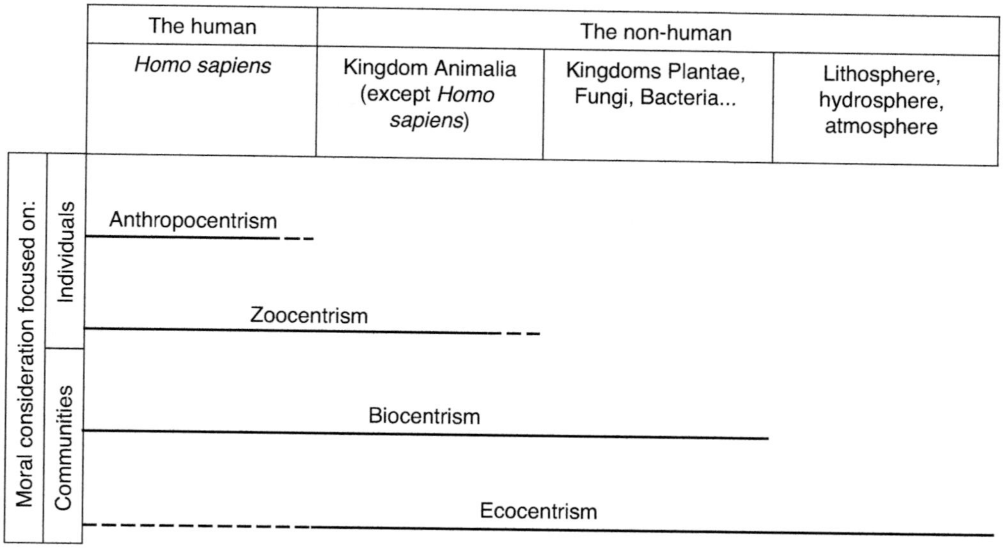

Fig. 2.1. Philosophical positions on moral consideration

(Steffen *et al.*, 2011). When human exceptionalism imposes itself on the intentionality of a total interspecific and eco-systemic equilibrium, an anthropocentric speciesism emerges to flirt with humanism (Cancino and De Alba, 2018; Anzoátegui, 2022).

The environmental imbalance provoked by humans has reached such a momentum that the notion of an Anthropocene as a new series or epoch in geologic time scale is rapidly gaining currency. However, zooethics does not see the annihilation of any sentient being (human or non-human) as a solution. The primary interests of all sentient beings are to preserve life and to avoid pain and suffering (Singer, 1987; Horta, 2011). Since, during the Anthropocene, humans have caused great suffering to all other animals, it is our moral obligation to reverse this situation. Moreover, though zooethics centres its attention on sentient beings, this doesn't mean that it is indifferent to the human impact on ecosystems. On the contrary, the value accorded to individual lives rests on a fundamental care for the environment. Thus, the promotion of a plant-based diet for humans, anti-natalism and the intervention on natural processes with positive results are some of the alternative measures to negotiate eco-systemic equilibrium with sentient forms of life (Horta, 2010; Faria, 2016).

Humanism rests on speciesism, a dogma of human superiority that allows for the reification of other animals and their appraisal as nothing more than tools for the satisfaction of human needs (Ryder, 1975; Lynn, 1998; Singer, 2009; Wolfe, 2010; Weitzenfeld and Joy, 2014; Horta, 2017; Kaltofen, 2018). The reification of animals goes hand in hand with the denial of their fundamental rights, their freedom and even their lives (Singer, 1975; Taylor and Signal, 2009). Speciesism relies on reason (Kaltofen, 2018), which since the Renaissance in Europe has been considered an exclusive faculty of humans bestowing them the arbitrary jurisdiction over everything that exists (Fig. 2.1).

If the ability to reason depends on having a brain, animals' inability to reason is an unfounded proposition. However, the issue of reason is irrelevant for an anti-speciesist and zoocentrist posthumanism grounded on sentience, that is, the awareness individuals have of their own pain and pleasure (Horta, 2007). Though in the last century philosophers like Regan (1983), Singer (1987) or Francione (1995) had already expounded on sentience, the solid scientific *The Cambridge Declaration on Consciousness* published in 2012, stating that 'humans are not unique in possessing the neurological substrates that generate consciousness [and that non-human] animals, including all mammals and birds, and many other creatures, including octopuses, also possess these neurological substrates' (Low, 2012, np), has advanced and reinforced the idea that animals are subjects of moral consideration. *The Toulon Declaration* of 2019, which states that all non-human animals should be considered legally and universally as persons and not objects and that their reification should be put to rest forever (Balmond *et al.*, 2019), is an important step in that direction.

When, from their position of privilege, humans systematically abuse the bodies and mental state of animals by forcing them to perform recreational and tourism activities, anti-speciesist posthumanism lights the alarm about the obvious fact that animals are being appraised as inferior, as mere objects (Horta, 2007). From this philosophical positioning, a different ethics that values animal welfare seeks to raise awareness about the suffering and damage inflicted on animals by exploitative work relations in several tourism activities. The legitimation of animals as persons defies anachronistic economic, social and legal structures that must change in order to establish fair human–animal relationships (Weitzenfeld and Joy, 2014).

Critical Approaches to Heritage: The Commodification of the Narrative of the Past

The critical analysis of heritage starts from the meaning of the term. In this chapter we shall consider the English and Spanish words for it. In English, there are two related concepts: patrimony refers specifically to the inheritance that, by means of a will, an individual receives via the paternal line, whereas heritage is the legacy – welcome or unwelcome – of one generation to another (Graham *et al.*, 2000). In Spanish, the word *patrimonio* serves both purposes and can be translated as patrimony or heritage. Feminist theory in the Spanish-speaking world maintains that the etymology of the word *patrimonio* enhances patriarchalism or the power of the patriarchs (Jiménez, 2016), that is, the interest of an existing elite in wielding power over public affairs. This elite deems itself superior to 'women, men, animals and ecosystems, whose productive and reproductive power it appropriates for its own benefit' (2014: 340; translation by the authors).

Though the words *patrimonio* and patrimony share the same root and are related to patriarchalism and its effects, this doesn't mean that heritage – presented usually as a neutral category – is free of the androcentric power relations affecting marginal groups (Lowenthal, 1998; Levy, 2013; Colella, 2018). In fact, Jiménez (2016) maintains that towers, belfries, obelisks, chimneys and other similar architectural forms are phallic metaphors that both materialize and perpetuate the dominant cultural heritage. In other words, heritage is prone to reproducing heteronormative and monolithic forms of understanding, naming and managing a legacy that serves the interests of those who proclaim it. In view of the intersectionality of gender, sexuality, race, age and ethnic origin, third-wave feminism calls for the need to deconstruct androcentric thinking (Levy, 2013), while anti-speciesist feminism (Faria, 2016) holds that such intersectionality should also include the category of species so that the interests of animals are equally protected.

Any human community could name what it considers its own heritage, but the mercantile logic imposes its preferences based on what a group of 'experts' considers legitimate on historical, scientific or artistic grounds (Roigé and Frigolé, 2010; Greffe, 2011). Thus the role of official institutions, from the local to the international, in naming, institutionalizing and preserving heritage is increasingly important. At the global level, the most widely recognized programmes in charge of managing heritage are UNESCO's World Heritage Convention and Intangible Cultural Heritage. The power of these programmes to unilaterally bestow heritage titles based on arbitrary and binary categories (such as natural versus cultural, tangible versus intangible) is questionable, as very often the sites or practices named as heritage are packaged together and turned into 'resources' for commercial exploitation (Tunbridge *et al.*, 2013).

Understood as a cultural product made of symbols that exalt a given identity and legitimize certain groups or societies (Roigé and Frigolé, 2010; Di Pietro *et al.*, 2018), heritage, it has been shown, tends to remain in former colonies long after the colonial powers have left, regardless of the ethical value of some of its contents (Tunbridge *et al.*, 2013). Bullfights are a case in point, for despite their prohibition in European countries they are still widely practised in Latin America.

From the institutional and academic spheres, heritage – whether natural or cultural – is usually presented to society as a positive construct under the argument that everything that is valuable for a group of people is valuable for all humanity (Dumoulin, 2002). But this pretended 'universalism' is at best relative, especially when associated with tourism, an activity that involves a wide array of actors. From a non-anthropocentric, anti-speciesist and posthumanist vision, heritage must be scrutinized in order to make sure it doesn't perpetuate a situation of subordination of one class of sentient beings (both human and non-human) to the interests of another (Aaltola, 2005). It is therefore important to identify the epistemological underpinnings of heritage, which include cultural (symbols, identity, representation), political (legitimation, authenticity, purity) and economic (resources, exchange value) aspects. The economic value of heritage is such that there is talk of a heritage economy – actively fostered by tourism – which commodifies ideas and cultural artefacts (Roigé and Frigolé, 2010).

The methodology of heritage naming must be revised, for enthusiastically and uncritically endorsing a traditional custom can have adverse consequences for the primary interests of human and non-human sentient beings. Picking a fragment of the past in order to exalt it and hold it as representative of tradition constitutes a gross simplification and conditioning of past and present realities (Lowenthal, 1998). And things get even worse when such fragment is reworked and reinterpreted from a contemporary perspective and put to new social uses (Roigé and Frigolé, 2010). Leaving aside the claims to objectivity of experimental and social sciences (Rakić, 2011), we maintain that a critical historical analysis supported by a systematic use of historiography has the potential to offer a good understanding of the complexity of the facts and constitutive processes of past scenarios (Lowenthal, 1998). In this regard, critics of heritage maintain that no heritage appointment should be exempt from being questioned, especially when it involves unethical practices (Jiménez, 2016).

Heritage has a strong connection with tourism – an unflagging consumer of commodified heritage elements – and the relationship between heritagization (the process of naming heritage) and touristification is undeniable (Nilsson, 2018). Heritagization is an engine for the production of ideas and cultural artefacts that suit the demands of the tourism market as well as political aspirations (Roigé and Frigolé, 2010; Nilsson, 2018). Driven by touristification, heritagization suddenly exalts some natural or cultural elements hitherto invisible for tourist operators, tourists and the receiving communities themselves.

Heritagization processes are usually heteronormative and based on a partial narrative of the past, the interests of a dominant group, a fictitious discourse of the authentic and a utilitarian conception of nature–society relations. Integrating the concern for animals' interests into the discussion is especially important, for heritage tourism has the potential to normalize many unethical practices. With heritagization, the bodies and experiences of other animals are often reified and commodified, reduced to symbols or objects of exotic consumption (Adams, 2020). Animals are generally portrayed as passive objects of desire or beauty, a situation that reinforces speciesist stereotypes and anthropocentric power imbalances (Salmen and Dhont, 2023).

The hegemonic narratives of masculinity are based on domination, control – and, eventually, real and symbolic violence – with power and authority as their desirable characteristics (Gallego, 2018). Influential humans impose this conception of the world on the general public and, in order to assert their power and control, resort to the exploitation of relatively vulnerable beings – human and non-human (Littlefield, 2010; Nurse, 2020). The separation of society from nature is essential for hegemonic masculinity to hold power and control over the environment (Twine, 2021). The binary vision of reality that sets natural heritage apart from cultural heritage legitimizes the exploitation and ill-treatment of animals, for when they are considered mere biological collectives their individual interests are nullified and their role is limited to providing environmental and other services to humans (Ćirović *et al.*, 2016; Alders *et al.*, 2021).

Ethical Issues on the Animals–Tourism Relationship: An Overview

In its report *Animal Heritage and Tourism*, the Foundation for Advice and Action in the Defence of Animals (FAADA, 2023) records several forms of abuse suffered by animals at the hands of tourism. It is estimated that globally at least 110 million people are involved annually in tourist activities affecting at least 550,000 animals. For example, the number of captive elephants used in tourism increased 70% between 2010 and 2020, and in South Africa alone there were 350 farms devoted to the breeding of lions in conditions of exploitation. In the Cayman Turtle Centre in the Cayman Islands, 95,000 turtles are bred for their meat in conditions of overcrowding, and in Petra, Jordan, 1300 horses, camels, mules and asses are exploited daily as a mode of transport. In Peru, tourists hold sloths every 5 minutes, and at the time of publication of this study, there were 3600 dolphins, orcas and belugas in 336 centres in 56 countries destined to recreational uses.

These are just a few examples evidencing the dire situations in which animals live in tourist destinations and that have elicited academic attention. Several years ago, Fennell (2012a) had noted the negligible number of tourism studies addressing ethical questions on the use of animals. More recently, Winter (2020) identified that between 2000 and 2019, 74 articles classified under the tourism–animals–ethics category had been published in ten English language journals of repute (six in the first five-year period, 12 in the second, 25 in the third and 31 in the fourth). Outside these journals, the author found another 75 texts. On that basis, Winter (2020) identified eight ethical positionings, five of them closer to anthropocentrism and the other three closer to post-anthropocentric, speciesist posthumanism (as discussed in the first section of this chapter).

Winter (2020) elaborates that: (i) from the anthropocentrism viewpoint, animals are regarded as instruments that reaffirm the superiority of humans (as in cynegetic tourism); (ii) utilitarianism risks justifying animal abuse on the grounds of a greater enjoyment by tourists (for example, the confinement of animals in zoos); (iii) welfarism justifies the painful death of animals as long as they have lived a 'good life', based on a subjective assessment of what constitutes fair treatment (as in the killing of animals to produce exotic dishes); (iv) in environmental ethics, the value accorded to the benefit of a species collective justifies the killing of a certain number of individuals of one of them, in total disregard for their interests (as happens in cynegetic tourism); and (v) the cultural aspects approach is framed in a cultural relativism that places cultural values and the traditional use of animals in tourism above the animals' primary interests.

As for the remaining three positionings, Winter (2020) refers to (i) ecofeminism's injunctions that the act of unravelling the patriarchal structures that legitimize the inequality between the sexes must be transferred to the relations of dominance of all animals by the human species, so as to foster interspecific empathy, the ethics of care and the abolition of speciesist practices in tourism such as the use of draught animals for transportation; (ii) the animal rights perspective, according to which humans are equal to other animals, since we all are 'subjects of a life' (Regan, 1983) with an intrinsic value expressed in physical and mental attributes (such as desires, memory, sense of the future) and thus any form of reification and commodification for entertainment, gastronomic, transportation or any other purpose in tourism is unethical and must be abolished; and (iii) the overall and multi-position discussions where it is assumed that binary categories such as human–animal, wild–domesticated are nothing but artificial borders to justify animal exploitation (Oliver, 2022). Winter (2020) emphasizes the importance of deconstructing categories that are taken as truths, as stated in the discussion of philosophic anti-speciesist posthumanism in the first section of this chapter.

Based on Leiper's (1979) tourism system, Markwell (2015) has presented a model of the ethical implications of the use of animals in tourism considering the unfolding of a tourist trip from the point of departure through transit to destination, including the pre-travel and post-travel stages.

During the pre-travel stage, potential tourists build their own representations of the animals from the experience of other tourists, mass media and social media, etc. These representations influence their desire to observe and consume animals as food or experience them as objects, or even reject them as pests that bring disease (Markwell, 2015). Contrastingly, humans almost never perceive themselves as dangerous vectors for animal disease, as happens in theme parks (Shani and Pizam, 2009). Often, animals that are considered heritage or part of heritage are treated as a tourist synecdoche, that is, the most important part of a whole without which the trip would not be completed (López and Quintero, 2021a). When this happens, the desire to have animals as gastronomic ingredients, souvenirs, ludic experiences, etc., is obvious.

During the post-travel stage, on the other hand, tourists may reinforce with their fellow interlocutors the representations of animals as objects for consumption through photographs, travel anecdotes, souvenirs made from animal parts, and other objects (Çalik and Çiftçi, 2013; Fennell, 2015), but when they become aware of the maltreatment to which many animals are subjected they may also try to re-elaborate their preconceived arguments so as to sensitize their peers on animal welfare.

The presence of animals in the tourist region of transit has been underexplored. Some animals are used as means of transport: horses, mules, donkeys, elephants, camels and dogs, among others, are regarded as exotic and 'docile' vehicles on account of their conditioning through feeding and the infliction of pain. They exist as much in cities (Taş Gürsoy, 2020), islands (Fennell, 2012a; Markwell, 2015), deserts and forests (Duffy and Moore, 2011) as in polar regions (Äijälä, 2019). But animals are also victims of tourist motor vehicles. As airplanes, trains and cars proceed, they leave behind dead or injured animals. Morris (2019) illustrates the case of kangaroos on Australian roads and Green and Giese (2004) examine the case of animals hit by trains or drawn into the blades and vortices of boats and airplanes. On the other hand, tourists engaged in air travel consider of the utmost importance the physical well-being of their company animals, exposed as they are to sudden changes in atmospheric pressure and temperature (Leggat and Speare, 2000; Carr and Cohen, 2009).

Markwell (2015) points to the fact that animal–tourist interactions are most evident at the destination, where animals perform as attractions or companions, are traded as commodities or used as photographic objects for image sharing or threat advisory. The interactions with these animals (except for those considered as threatening) are usually valued as positive and gratifying, even if for the animal in question they often prove to be harmful, painful and cruel experiences (Reynolds and Braithwaite, 2001; Fennell, 2012a; Markwell, 2015). The fact that tourists have their experiences with animals in ludic, festive or relaxed environments contributes to the latter's reification and to the obliteration and normalization of their suffering (Markwell, 2015; Carr and Broom, 2018; Kline, 2018; Rickly and Kline, 2021).

As critical academic scrutiny has revealed, under the protection of categories such as nature tourism, ecotourism, sports tourism, sustainable tourism, cultural tourism, heritage tourism and gastronomic tourism the lives of animals are systematically aggrieved, be it when they are chased for hunting (Tickle and von Essen, 2020), simply observed in the wild (Moorhouse *et al.*, 2017), confined to be admired in zoos and aquariums (Frost, 2011), bred in order to later being hurt or killed in traditional festivities (Acharya and Wilson, 2020), caged for their sale in traditional markets (Whiting *et al.*, 2013), forcibly trained to perform pirouettes in aquatic parks and circuses (Massey and Randriamarohaja, 2019), force-fed to obtain their meat, secretions or excreta used in exotic gastronomies (Kline, 2018) or perform as sexual objects (Panko and George, 2018). Some of these topics are approached in this book.

Tourism, Animals and Heritage: A Complex Interaction

It has been argued that heritage can be named by any social entity, from a small local group to a specialized institution of regional, national or global reach. However, in an attempt to offer a general overview of the interaction between heritage and animals, we have decided to focus on the UNESCO designations. At the time of conducting our study, the World Heritage Convention had 1157 sites registered as natural, mixed and cultural heritage (UNESCO, 2023a) and 676 practices and cultural expressions as Intangible Cultural Heritage (UNESCO, 2023b), which makes for a total of 1833 entries. From these lists, we selected those sites, practices and heritage expressions in which animals are involved. This information allowed us to construct the tables presented below as well as to develop the main arguments of this section.

The World Heritage Convention website (UNESCO, 2023a) has a page for each heritage site with several tabs. We focused on the 'Description' tab, which offers an introductory summary on the relevance of the heritage site followed by a photographic mosaic and a section called 'Outstanding Universal Value'. Since our interest lay in detecting the relevance of animals in the site's definition rather than scrutinizing the whole section for each heritage site, we looked for any mention of specific animal species or the words 'animal' or 'fauna' and added the relevant entry to our list. In the absence of a mention of animals, we looked for related terms such as 'biodiversity', 'wilderness', 'environment' or 'ecosystem' and looked at the 'Outstanding Universal Value' section for any direct or indirect mention of animals, in order to integrate the conforming entry into our list. We left out entries with images of animals in the photographic mosaic but no mention of them in the text. We also left out those sites in which the allusion was not to animals themselves but to their artistic representations, unless they were physically present. We thus built two lists, one for natural heritage, with 154 sites (Table 2.1) and another one for mixed and cultural heritage, with 39 sites (Table 2.2). In total, we listed 193 sites.

As for Intangible Cultural Heritage, we went through each nomination and proceeded in the same way, leaving out the artistic representations of animals. We obtained 60 practices and cultural expressions related to animals (Table 2.3).

The interaction between heritage sites, tourism, ecosystems and animal well-being leads to the formation of contested spaces (McLennan *et al.*, 2017). The commodification of spatial elements by tourism goes against the interests of the preservation of natural environments and the basic interests of animals. The World Heritage Convention's standard narrative to justify the designation of natural sites as heritage is their ecological-scenic value and not the presence of animals *per se*, which belies an ecocentric perspective. Moreover, two forms of speciesism are evident: one having to do with human self-appreciation *vis-à-vis* other animal species, and the other with the appraisal that humans make of the latter and the moral value accorded to each one of them.

The first kind of speciesism is evident in the very concept of *human* heritage, a unidirectional vision intending to preserve animal species – regarded as an 'accessory' part of ecosystems – in order to perpetuate the privileges of present and future human societies, as the concept of sustainable development goes (Srinivasan and Kasturirangan, 2017; Sterling, 2020). The second kind is evident when humans differentiate 'emblematic' and 'charismatic' animal species from the rest (Hosey *et al.*, 2020). The first ones are mostly vertebrate species employed as tourist attractions that are present in the conservation plans and generate income for local economies (Skibins *et al.*, 2013). Conversely, non emblematic and non charismatic species are destined for consumption, experimentation or labelled as 'pests' to be controlled through biopower (Büscher, 2018) and therefore obtain less empathy from tourists, unless they are part of a specific heritage (Skibins *et al.*, 2017).

Tourism at terrestrial UNESCO heritage sites alters the natural development processes even when activities are considered 'low impact', such as watching animals in the wild. Noise disturbance is a big problem associated with this activity, as frightened adult animals tend to flee or hide, leaving their eggs or young at the mercy of all kinds of events. Pollution is also a problem (Fennell, 2012a). Since wild animals avoid humans, the tactics to attract them through sound imitation and the strategic placing of food, or visitors' access to reproduction sites (Maciejewski and Kerley, 2014), interfere in the very environment the heritage nomination intends to preserve, especially in conditions of 'touristification'.

Among natural heritage sites, coastal and maritime environments are probably the most fragile. The use of vessels by the local population

Table 2.1. World Heritage Convention: natural sites where animals are involved. Based on UNESCO (2023a).

Country	World Heritage Convention site[a]	In reference to animals	Country	World Heritage Convention site[a]	In reference to animals
Argentina	Los Alerces National Park	Endemic and threatened fauna	Japan	Shiretoko	Blackiston's fish owls, Steller's sea lions
Argentina	Península Valdés	Marine mammals	Japan	Shirakami-Sanchi	Black bears, serow, birds
Argentina + 1	Iguazu National Park	Tapirs, giant anteaters, howler monkeys	Kazakhstan	Saryarka – Steppe and Lakes…	Siberian white cranes, pelicans
Australia	Ningaloo Coast	Whale sharks, sea turtles	Kenya	Kenya Lake System in the Great Rift…	Birds, mammals
Australia	Heard and McDonald Islands	Subantarctic fauna	Kenya	Mount Kenya National Park…	African elephants
Australia	Shark Bay, Western Australia	Endangered mammals	Kenya	Lake Turkana National Parks	Nile crocodiles, hippopotamuses, snakes
Australia	Wet Tropics of Queensland	Marsupials, singing birds	Kiribati	Phoenix Islands Protected Area	Corals, fish, marine mammals, birds
Australia	Lord Howe Island Group	Birds	Madagascar	Rainforests of the Atsinanana	Primates, lemurs
Australia	Great Barrier Reef	Corals, fish, molluscs, dugong	Madagascar	Tsingy de Bemaraha Strict Nature…	Lemurs, birds
Bangladesh	The Sundarbans	Bengal tigers, estuarine crocodiles	Malawi	Lake Malawi National Park	Fish
Belarus + 1	Białowieża Forest	European bison	Malaysia	Kinabalu Park	Mammals, birds, amphibians
Belize	Belize Barrier Reef Reserve System	Marine turtles, manatees, crocodiles	Malaysia	Gunung Mulu National Park	Swiftlets, bats
Benin + 2	W-Arly-Pendjari Complex	Large mammals	Mauritania	Banc d'Arguin National Park	Turtles, dolphins

Continued

Table 2.1. Continued

Country	World Heritage Convention site[a]	In reference to animals	Country	World Heritage Convention site[a]	In reference to animals
Bolivia	Noel Kempff Mercado National Park	Birds, vertebrates	Mexico	Archipiélago de Revillagigedo	Seabirds, manta rays, whales, dolphins
Botswana	Okavango Delta	Large mammals	Mexico	El Pinacate and Gran Desierto de Altar…	Freshwater fish, Sonoran pronghorns
Brazil	Brazilian Atlantic Islands…	Fish, turtles, marine mammals, seabirds	Mexico	Monarch Butterfly Biosphere Reserve	Monarch butterfly
Brazil	Cerrado Protected Areas…	Large mammals, birds	Mexico	Islands and Protected Areas […] Gulf of…	Marine mammals
Brazil	Central Amazon Conservation…	Arapaima fish, manatees, river dolphins	Mexico	Whale Sanctuary of El Vizcaino	Grey whale
Brazil	Pantanal Conservation Area	Mammals, birds, reptiles, fish	Mexico	Sian Ka'an	Birds, terrestrial vertebrates
Brazil	Discovery Coast Atlantic Forest…	Mammals, birds, amphibians, reptiles	Mongolia + 1	Landscapes of Dauria	Birds, Mongolian gazelles
Brazil	Atlantic Forest South-East Reserves	Mammals, birds	Mongolia + 1	Uvs Nuur Basin	Birds, gerbils, jerboas, marbled polecats
Bulgaria	Srebarna Nature Reserve	Birds	Namibia	Namib Sand Sea	Invertebrates, reptiles, mammals
Cameroon	Dja Faunal Reserve	Primates	Nepal	Chitwan National Park	Asiatic rhinoceroses, Bengal tigers
Cameroon + 2	Sangha Trinational	Nile crocodiles, goliath tigerfish	Nepal	Sagarmatha National Park	Snow leopards, lesser pandas
Canada	Canadian Rocky Mountain Parks	Soft-bodied marine animals	New Zealand	New Zealand Sub-Antarctic Islands	Pelagic seabirds, penguins
Canada	Wood Buffalo National Park	Wild bison, whooping cranes	New Zealand	Te Wahipounamu–South West New…	Alpine parrot, flightless bird

Continued

Table 2.1. Continued

Country	World Heritage Convention site[a]	In reference to animals	Country	World Heritage Convention site[a]	In reference to animals
Canada	Nahanni National Park	Wolves, grizzly bears, caribou	Niger	Air and Ténéré Natural Reserves	Wild animals
Canada + 1	Waterton Glacier International Peace…	Mammals	Norway	West Norwegian Fjords–Geirangerfjord…	Marine mammals
Canada + 1	Kluane/Wrangell-St Elias/ Glacier…	Grizzly bears, caribou, Dall's sheep	Panama	Coiba National Park and […] Marine…	Eagles, pelagic fish, marine mammals
Central African Republic	Manovo-Gounda St Floris National…	Mammals, waterfowl	Panama	Darien National Park	Remarkable wildlife
Chad	Lakes of Ounianga	Fish	Peru	Manú National Park	Birds, giant otters, giant armadillos, jaguars
China	Migratory Bird Sanctuaries […] Yellow…	Birds	Peru	Huascarán National Park	Spectacled bears, Andean condors
China	Fanjingshan	Monkeys, salamanders, Reeve's pheasants	Philippines	Mount Hamiguitan Range Wildlife…	Philippine eagles, Philippine cockatoos
China	Qinghai Hoh Xil	Tibetan antelope	Philippines	Tubbataha Reefs Natural Park	Birds, marine turtles
China	Hubei Shennongjia	Salamanders, monkeys, leopards	Portugal	Laurisilva of Madeira	Madeiran long-toed pigeons
China	Sichuan Giant Panda Sanctuaries…	Giant pandas	R. of Korea	Getbol, Korean Tidal Flats	Birds, octopuses, crabs, polychaetes
China	Three Parallel Rivers of Yunnan…	Endangered animals	Romania	Danube Delta	Birds, fish
China	Huanglong Scenic and Historic…	Giant pandas, golden snub-nosed monkeys	Russia	Putorana Plateau	Wild reindeer
China	Wulingyuan Scenic and Historic…	Asiatic black bears, Chinese water deer	Russia	Natural System of Wrangel Island…	Walruses, polar bears, grey whales, birds

Continued

Table 2.1. Continued

Country	World Heritage Convention site[a]	In reference to animals	Country	World Heritage Convention site[a]	In reference to animals
Colombia	Malpelo Fauna and Flora Sanctuary	Sharks, giant groupers, billfish	Russia	Central Sikhote-Alin	Tigers, brown bears, lynx, musk deer
Colombia	Los Katíos National Park	Threatened animal species	Russia	Western Caucasus	European bison
Costa Rica	Área de Conservación Guanacaste	Endangered animals	Russia	Golden Mountains of Altai	Snow leopards
Costa Rica	Cocos Island National Park	Sharks, rays, tuna, dolphins	Russia	Lake Baikal	Unusual freshwater faunas
Côte d'Ivoire	Taï National Park	Pygmy hippopotamuses, monkeys	Russia	Volcanoes of Kamchatka	Fish, sea otters, brown bears
Côte d'Ivoire + 1	Mount Nimba Strict Nature Reserve	Viviparous toads, chimpanzees	Saint Lucia	Pitons Management Area	Finfish, corals, birds, rodents, bats
Croatia	Plitvice Lakes National Park	Bears, wolves, rare birds	Senegal	Djoudj National Bird Sanctuary	Pelicans, herons, egrets, cormorants
Cuba	Alejandro de Humboldt National Park	Vertebrates and invertebrates	Senegal	Niokolo-Koba National Park	Chimpanzees, lions, leopards, elephants
D.R. Congo	Okapi Wildlife Reserve	Primates, birds, okapis	Seychelles	Aldabra Atoll	Giant tortoises
D.R. Congo	Salonga National Park	Chimpanzees, elephants, 'false' crocodiles	Solomon Islands	East Rennell	Corals
D.R. Congo	Garamba National Park	Hippos, gorillas, birds	South Africa	iSimangaliso Wetland Park	Turtles, whales, dolphins, whale-sharks
D.R. Congo	Kahuzi-Biega National Park	Gorillas	Spain	Doñana National Park	Birds
D.R. Congo	Virunga National Park	Hippos, gorillas, birds	Sri Lanka	Central Highlands of Sri Lanka	Langurs, slender lorises
Denmark + 2	Wadden Sea	Harbour seals, grey seals, harbour porpoises	Sri Lanka	Sinharaja Forest Reserve	Birds, mammals, insects, reptiles

Continued

Table 2.1. Continued

Country	World Heritage Convention site[a]	In reference to animals	Country	World Heritage Convention site[a]	In reference to animals
Dominica	Morne Trois Pitons National Park	Reptiles, amphibians, birds	Sudan	Sanganeb Marine National Park…	Seabirds, mammals, turtles
Ecuador	Sangay National Park	Mountain tapirs, Andean condors	Suriname	Central Suriname Nature Reserve	Jaguars, armadillos, otters, tapirs, primates
Ecuador	Galápagos Islands	Land iguanas, giant tortoises, finches	Tajikistan	Tajik National Park…	Birds, sheep, snow leopards, ibex
Ethiopia	Simien National Park	Gelada baboons, Simien foxes	Tanzania	Kilimanjaro National Park	Mammals
France	French Austral Lands and Seas	King penguins, albatrosses	Tanzania	Selous Game Reserve	Elephants, rhinoceroses, cheetahs
France	Lagoons of New Caledonia…	Turtles, whales, dugongs	Tanzania	Serengeti National Park	Wildebeest, gazelles, zebras
France	Gulf of Porto: Calanche of Piana…	Seagulls, cormorants, sea eagles	Thailand	Kaeng Krachan Forest Complex	Birds, crocodiles, wild dogs, banteng
Gabon	Ivindo National Park	Fish, crocodiles, butterflies, mammals	Thailand	Dong Phayayen-Khao Yai Forest…	Birds, reptiles, amphibians, mammals
Georgia	Colchic Rainforests and Wetlands	Vertebrates and invertebrates	Thailand	Thungyai-Huai Kha Khaeng Wildlife…	Elephants, tigers, birds
Honduras	Río Plátano Biosphere Reserve	Abundant wildlife	Tunisia	Ichkeul National Park	Ducks, geese, storks, pink flamingos
Iceland	Surtsey	Birds and invertebrates	Uganda	Bwindi Impenetrable National Park	Birds, butterflies, mountain gorillas
India	Great Himalayan National Park…	Threatened fauna	Uganda	Rwenzori Mountains National Park	Birds, elephants, chimpanzees, monkeys
India	Western Ghats	Birds, amphibians, reptiles, fish	UK	Gough and Inaccessible Islands	Gallinule, gough rowettie, invertebrates

Continued

Table 2.1. Continued

Country	World Heritage Convention site[a]	In reference to animals	Country	World Heritage Convention site[a]	In reference to animals
India	Nanda Devi and Valley of Flowers…	Black bears, snow leopards	UK	Henderson Island	Birds
India	Sundarbans National Park	Tigers, aquatic mammals, birds, reptiles	USA	Hawaii Volcanoes National Park	Birds
India	Kaziranga National Park	Many mammals	USA	Yosemite National Park	Variety of fauna
India	Keoladeo National Park	Aquatic birds	USA	Great Smoky Mountains National Park	Salamanders
Indonesia	Tropical Rainforest Heritage of Sumatra	Sumatran orangutans	USA	Mammoth Cave National Park	Variety of fauna
Indonesia	Lorentz National Park	Birds, mammals	USA	Olympic National Park	Anadromous fish, northern spotted owls
Indonesia	Ujung Kulon National Park	Endangered animals	USA	Everglades National Park	Birds, reptiles, manatees
Indonesia	Komodo National Park	Komodo dragons	USA	Yellowstone National Park	Grizzly bears, wolves, bison, wapitis
Iran	Hyrcanian Forests	Birds, mammals	Viet Nam	Phong Nha-Ke Bang National Park	Animal endemic species
Japan	Amami-Oshima Island, Tokunoshima…	Amami rabbits, long-haired rats	Yemen	Socotra Archipelago	Reptile, land snails, birds, corals
Japan	Ogasawara Islands	Bonin flying foxes, fish, cetaceans, corals	Zimbabwe	Mana Pools National Park…	Elephants, buffaloes, leopards

[a]Some of the names of sites are not given in full due to space constraints.

Table 2.2. World Heritage Convention: mixed and cultural sites where animals are involved. Based on UNESCO (2023a).

Country	World Heritage Convention site (Mixed)[a]	In reference to animals	Country	World Heritage Convention site (Cultural)[a]	In reference to animals
Albania + 1	Natural and Cultural Heritage of the Ohrid	Freshwater fauna	Andorra	Madriu-Perafita-Claror Valley	Pastoral societies
Australia	Kakadu National Park	Endemic animals	Bahrain	Pearling, Testimony of an Island Economy	Oyster beds
Brazil	Paraty and Ilha Grande – Culture…	Jaguars, peccaries, primate species	Canada	Red Bay Basque Whaling Station	Pre-industrial whaling station
Canada	Pimachiowin Aki	Fishing and hunting	Canada	Head-Smashed-In Buffalo Jump	American bison skeletons
China	Mount Wuyi	Reptile, amphibian and insect species	Czechia	Landscape for Breeding and […] Horses	Horses
Colombia	Chiribiquete National Park…	Fauna endemism	Denmark	Aasivissuit – Nipisat. Inuit Hunting…	Hunting of land and sea animals
France + 1	Pyrénées – Mont Perdu	Pastoral societies	Denmark	Kujataa Greenland: Norse and Inuit Farming…	Historic grazing, marine mammal hunting
Iraq	The Ahwar of Southern Iraq: […] Biodiversity…	Birds, mammals, fish, reptiles	Denmark	The par force hunting […] in North Zealand	Hunting extinct with hounds
Lesotho + 1	Maloti-Drakensberg Park	Cape vultures, bearded vultures	France	Taputapuātea	Coral reef
Mexico	Ancient Maya City […] Forests of Calakmul…	Mammals, butterflies, parrots, bats	France	The Causses and the Cévennes…	Pastoral societies
Palau	Rock Islands Southern Lagoon	Birds, dugongs, sharks	Hungary	Hortobágy National Park	Pastoral societies
Peru	Río Abiseo National Park	Yellow-tailed woolly monkeys	Iran	Cultural Landscape of Hawraman/Uramanat	Pastoral societies
Sweden	Laponian Area	Movement of livestock (reindeer)	Iran	Cultural Landscape of Maymand	Pastoral societies
Tanzania	Ngorongoro Conservation Area	Wildlife, pastoral societies	Japan	Tomioka Silk Mill and Related Sites	Silkworm

Continued

Table 2.2. Continued

Country	World Heritage Convention site (Mixed)[a]	In reference to animals	Country	World Heritage Convention site (Cultural)[a]	In reference to animals
UK	St Kilda	Puffins, gannets	Kenya	Thimlich Ohinga Archaeological Site	Pastoral societies
USA	Papahānaumokuākea	Coral reefs	Mongolia	Orkhon Valley Cultural Landscape	Pastoral societies
			Norway	Vegaøyan – The Vega Archipelago	Fishing, duck nesting
			Senegal	Bassari Country: Bassari, Fula and Bedik…	Pastoral societies
			Senegal	Saloum Delta	Fishing, shellfish gathering
			South Africa	Richtersveld Cultural and Botanical…	Pastoral societies
			UK	The English Lake District	Pastoral societies
			UK	Gorham's Cave Complex	Traces of hunting birds, marine animals
			Uruguay	Fray Bentos Industrial Landscape	Historic meat food chain

[a]Some of the names of sites are not given in full due to space constraints.

Table 2.3. Intangible Cultural Heritage where animals are involved. Based on UNESCO (2023b).

No.	Year	Country	Intangible cultural heritage of humanity[a]	Description emphasizing animals
1	2022	Afghanistan + 6	Sericulture, production of silk for weaving	Silkworms and cocoons production
2	2022	Austria + 7	Lipizzan horse breeding traditions	Horse breeding to be used in events
3	2022	Jordan	Al-Mansaf in Jordan, a festive banquet…	Chunks of sheep or goat meat with yogurt sauce
4	2022	North Korea	Pyongyang Raengmyon custom	Noodles with meat, among other things
5	2022	Saudi Arabia + 2	Alheda'a, oral traditions of calling camel flocks	Music played by herders to camels. Herders train their camels
6	2021	Italy	Truffle hunting and extraction…	Carried out with the help of a trained dog
7	2021	Morocco	Tbourida	Equestrian performance
8	2021	Senegal	Ceebu Jën, a culinary art of Senegal	Dish made with fish and molluscs, among other things
9	2021	Slovenia	Beekeeping in Slovenia, a way of life	Beekeeping for food and traditional medicine
10	2021	United Arab Emirates + 23	Falconry, a living human heritage	Training and flying falcons and other birds of prey
11	2020	Poland + 1	Tree beekeeping culture	Beliefs connected to wild bees breeding in tree hives
12	2020	Spain	Wine horses	The equestrian ritual with dressing the horses
13	2020	Tunisia	Charfia fishing in the Kerkennah Islands	Local fish
14	2020	United Arab Emirates + 1	Camel racing, a social practice and a festive…	Camel racing
15	2019	Austria + 2	Transhumance, the seasonal droving of livestock…	Animals are driven by herders with their dogs and horses
16	2019	Belarus	Spring rite of Juraŭski Karahod	Ritual: Animals are led out of the barn after winter
17	2019	Mongolia	Dish made with fish and molluscs, among other things	Fermented beverage made from mare's milk in cowhide vessel
18	2019	Türkiye	Traditional Turkish archery	Equipment made with horns, tendons, silk and leather
19	2018	Kazakhstan	Traditional […] rites of the Kazakh horse breeders	Separation of mares and foals from the herds

Continued

Table 2.3. Continued

No.	Year	Country	Intangible cultural heritage of humanity[a]	Description emphasizing animals
20	2018	Kenya	Enkipaata, Eunoto and Olng'esherr, three male rites	One of the three rites is to eat meat
21	2018	Oman	Horse and camel Ardhah	Shows by horse and camel riders
22	2018	Zambia	Mooba dance of the Lenje ethnic group…	Whisk made out of an animal's tail
23	2017	Iran	Chogān, a horse-riding game…	Horse riding
24	2017	Kazakhstan	Kazakh traditional Assyk games	Game of talus bone of a sheep
25	2017	Kyrgyzstan	Kok boru, traditional horse game	Teams on horseback; players manoeuvre a goat's carcass
26	2016	Mexico	Charrería, equestrian tradition in Mexico	Roping and reining using mares and bulls, on foot or horseback
27	2016	Nigeria	Argungu international fishing and cultural festival	Ater competitions: hand fishing, wild duck catching
28	2016	Tajikistan	Oshi Palav, a traditional meal…	Vegetables, rice, meat and spices
29	2016	Uzbekistan	Palov culture and tradition	Rice, meat, spices and vegetables
30	2015	Austria	Classical horsemanship and the High School…	Classical horsemanship
31	2015	Mongolia	Coaxing ritual for camels	Encourage a female camel to accept a new-born calf
32	2015	Portugal	Manufacture of cowbells	Locate and control the livestock (soundscape)
33	2015	Turkmenistan	Epic art of Gorogly	Scenic oral tradition as oral transmission of stock breeding
34	2014	Mongolia	Mongolian knuckle-bone shooting	Bones of their livestock are used in religious rites
35	2013	Algeria	Annual pilgrimage to the mausoleum…	Equestrian competitions
36	2013	Azerbaijan	Chovqan, a traditional Karabakh horse-riding game…	Horse-riding game
37	2013	Belgium	Shrimp fishing on horseback in Oostduinkerke	Shrimp fishing
38	2013	Cyprus + 6	Mediterranean diet	Fishing, animal husbandry

Continued

Table 2.3. Continued

No.	Year	Country	Intangible cultural heritage of humanity[a]	Description emphasizing animals
39	2012	United Arab Emirates + 1	Al-Taghrooda, traditional Bedouin chanted poetry…	Camel races
40	2011	Brazil	Yaokwa, the Enawene Nawe people's ritual…	Fish and prepare offerings with fish
41	2011	Czechia	Ride of the Kings…	The entourage rides on decorated horses
42	2011	France	Equitation in the French tradition	School of horseback riding
43	2011	Japan	Mibu no Hana Taue, ritual of transplanting rice…	Cattle dressed with saddles and necklaces
44	2011	Türkiye	Ceremonial Keşkek tradition	Chunks of meat on the bone
45	2011	United Arab Emirates	Al Sadu, traditional weaving skills…	Shear sheep, camels and goats and clean wool
46	2010	Belgium	Houtem Jaarmarkt […] fair and livestock market…	Livestock market. Seeing, touching and buying animals
47	2010	Croatia	Sinjska Alka, a knights' tournament in Sinj	Horse riding
48	2010	France	Gastronomic meal of the French	Courses that include fish, meat, among others
49	2010	Japan	Yuki-tsumugi, silk fabric production technique	Silkworms
50	2010	Mexico	Traditional Mexican cuisine…	Animal meat used for dishes
51	2009	China	Sericulture and silk craftsmanship of China	Silkworms
52	2009	Mali	Sanké mon, collective fishing rite of the Sanké	The rite begins with the sacrifice of roosters, goats
53	2008	Cambodia	Sbek Thom, Khmer shadow theatre	Puppets are made from a single piece of leather
54	2008	Costa Rica	Oxherding and oxcart traditions in Costa Rica	Oxherding and oxcart
55	2008	Indonesia	Wayang puppet theatre	Flat leather shadow puppet
56	2008	Jordan	Cultural space of the Bedu in Petra and Wadi Rum	Camel husbandry, pastoral culture
57	2008	Mali	Cultural space of the Yaaral and Degal	Competitions about decorated herd

Continued

Table 2.3. Continued

No.	Year	Country	Intangible cultural heritage of humanity[a]	Description emphasizing animals
58	2008	Mongolia	Traditional music of the Morin Khuur	Soundboard with animal skin; strings and bow with horsehair
59	2008	Morocco	Moussem of Tan-Tan	Camel and horse-breeding competitions
60	2008	Uzbekistan	Cultural space of Boysun District	Goat fights and horse races

[a]Some entries in this column are not given in full due to space constraints.

affects the lives of other animals, including sea mammals and fish. Particularly vulnerable to the presence of tourists are reef areas. Some examples are the Australian Coral Reef and the Mesoamerican Reef in Belize (Diedrich, 2007). In some sites, fishing is allowed, while in others it is prohibited. Diving, however, is generally allowed, even though the degradation of ecosystems by the very presence of divers, their interaction with marine life, and the collection of shells and other organisms has been documented (Doiron and Weissenberger, 2014; Giglio *et al.*, 2020).

In coastal areas in which birds and marine mammals transit, nest and reproduce (such as deltas and mangroves), the watching of animals in the wild is a highly disturbing tourist activity. At the Danube Delta in Romania, for example, it has been found that the noise caused by motor vessels upsets the feeding, resting, nesting and migration patterns of birds (Gâştescu and Toma, 2019), which end up abandoning their habitat. Vessel trails can also affect the quality of the water and the food resources of birds and other animals.

One of the most ethically questionable practices in many of the World Heritage Convention sites (Tables 2.1 and 2.2) and Intangible Cultural Heritage sites (Table 2.3) is cynegetic tourism (Fletcher-Barnes, 2020). As discussed in the previous sections, the hunting of animals for recreation is closely associated with the patriarchal dominion over nature and the feminine, as well as with violent and predatory intercourse (Luke, 1998). Though hunting is often justified under the apothegm 'kill to preserve', from the perspective of a zoocentric philosophical posthumanism the practice of deliberately hounding and killing or injuring animals with weapons has no ethical standing, and the trading and dismembering of animals to display them as trophies in the name of conservation and heritage is an oxymoron.

In the apocalyptic scenario painted by the Anthropocene, the narrative that heritage sites (natural, cultural or mixed) constitute the pristine or authentic 'last redoubts' on the earth's surface lends them an extraordinary value and compels tourists to visit them before it's too late (Tucker and Shelton, 2014). But are we expected to accept that the best way to preserve such sites is to turn them into heritage items for tourist consumption?

Through a process of ontologization (Ávila, 2014; Fleisner, 2019), animals are assigned what humans consider is their 'true' nature. Their behaviour, anatomy and sex are anthropomorphized (Nielsen, 2008) to suit human comprehension. In a tacit or explicit objectification of domesticated and wild animals, the World Heritage and the Intangible Cultural Heritage registries classify animals as mixed and cultural heritage (Tables 2.2 and 2.3). The 'animals for tourism' (Kline, 2023) and 'animals of heritage' (Orlić and Marjanić, 2018) are thus considered as 'seasoning' elements of cultural environments, or as herding animals, hunting and fishing animals, meat-and-milk-producing animals, pearl-producing animals, silk-producing animals, animals as mementoes, companion animals, animals for sustenance, animals as means of transportation of human bodies or merchandise, animals as performers at fairs, festivities and traditional spectacles – in which they must run, make pirouettes or be abused – animals as ingredients for exotic dishes, and so on.

In a large number of mixed and cultural heritage sites of both modern and peasant societies

(Tables 2.2 and 2.3) animals in total or partial confinement are destined to transportation, consumption or both. What these animals have in common is that, within a capitalist framework, they are nothing but slaves that can be conveniently traded as private property (Hribal, 2003). Camels, cattle, pigs, geese, ducks, chickens and others have been part of the working class – that has contributed to the accumulation of wealth (Hribal, 2003) without any remuneration, in the same way as adult and infant humans have lived under conditions of exploitation since the Industrial Revolution, aggravated today by the demands of heritage tourism.

Enormous amounts of food and other products of animal origin are part and parcel of heritage nominations in different parts of the world. This reinforces the idea that animals are the natural source of food and resources for industrial development in capitalist societies. The Fray Bentos Industrial Landscape in Uruguay is an example of the many landscapes of large-scale meat production throughout the world, especially in South America, where the presence of *asados*, *rodizios* and other meat specialties have become important tourist attractions. In various episodes of human history, animal species have been depleted to the point of extinction, as happened at Red Bay Basque Whaling Station in Canada where during the 16th century the massive extraction of oil from blubber to be sent to Europe for lighting made the whale population practically extinct (UNESCO, 2023b). Outside of capitalist exploitation, an example of animal maltreatment for human advantage is the ancient practice of driving bison herds over a precipice in order to kill them easily at the Head-Smashed-In Buffalo Jump, Canada (UNESCO, 2023b).

The gourmet culinary heritage based on animal ingredients effectively hides the life and death realities of animals. For example, *foie gras*, considered a delicacy in French gastronomy, is obtained by force-feeding geese and ducks to the extent that their liver is hypertrophied. Making meat production into cultural heritage not only legitimizes a massive productive process based on the confinement and slaughter of millions of animals belonging to different species but it also encourages tourists to visit restaurants whose menus are based on meat and its derivatives. This perpetuates the idea of carnism (Joy, 2010), a belief system that justifies the consumption of animal products at the expense of the suffering of sentient beings and the environmental effects caused by the industry.

Behind the festive, ludic or formal atmospheres of multiple recreational heritage events oblivious of animal interests hides animal suffering. Large parades, races of festooned camels and horses, oxen pulling profusely embellished carts are some of the practices that directly hurt animals' basic interests. Many such practices are listed in Tables 2.2 and 2.3, which are anything but exhaustive for, outside the UNESCO list, there are hundreds of other practices in which animal suffering is related to heritage. Bullfights and cockfights are classical examples of this. In an effort to describe this complex reality, Quintero and López (2018) and López and Quintero (2021b) have coined the term 'animal dark tourism' to account for the fact that the tourist attraction of such practices resides in the slaughter and disposal of animals.

Conclusions

The purpose of this chapter has been to contribute to the academic discussion on the commodification of animals in heritage sites of interest to tourists from the perspective of a post-anthropocentric and anti-speciesist philosophical posthumanism. The deconstruction of heritage sites and practices that reinforce patriarchal values in tourism must be accompanied by the 'de-ontologization' of the animal category, an exercise that reveals animals in their true light as beings whose primary interests go contrary to the tourist practices to which they are tied (Ávila, 2014). This would ensure the recognition of animals as individuals with their own needs and rights and challenge the policies and practices that thrive at their expense. It is therefore necessary to redesign the precepts of heritage from a critical perspective and accept that what suits some humans doesn't necessarily suit the rest of humanity and the animal kingdom.

In most cases in which animals are part of natural, cultural or mixed heritage according to the UNESCO World Heritage Convention and Intangible Cultural Heritage, the lives of animals in the wild or in confinement are at risk, for tourism is a growing impulse demanding their services and their bodies. Tourists–animal relations usually evidence unethical practices that may or not be voluntary (Reynolds and Braithwaite, 2001; Fennell, 2012b; Markwell, 2015). Being involved in situations that are alien to them physically affects their bodies, their physiology and their emotions (Vint, 2010).

Tourism studies must raise awareness of the fact that animals are at risk in heritage sites. They must produce information on the ways of life of animals, their primary interests and their exposure to unethical practices. They must offer the necessary evidence to support the arguments in favour of modifying laws and public policies so that animals can be protected in existing and future heritage sites. Likewise, heritage entities must devise supervision mechanisms to prevent and punish animal cruelty. This obviously implies challenging the economic, social and legal structures that perpetuate the exploitation of animals, and anti-speciesist philosophical perspectives can provide the orientation.

In writing this text we realized that in heritage tourism environments liminal, stray and street animals are invisible, as is often the abuse and mistreatment they suffer (Perez, 2020). Focused as they are on economic gain, tourist policies have no place for these irrelevant animals, uncomfortable elements of the heritage tourist landscape. However, many tourists are aware of the dire situation of stray animals and in some cases they have either adopted a few, reported animal abuse situations or paid for the treatment expenses of sick or injured animals.

Acknowledgements

This chapter is derived from the UNAM-PAPIIT IN30362 research project: *Confluencia de las geografías de los animales y del turismo desde la ética poshumanista: una perspectiva multidisciplinaria desde y sobre Latinoamérica* (Confluence of both animal and tourism geographies from a posthumanist ethics: a multidisciplinary perspective from and about Latin America). We acknowledge the support provided by José Alejandro Garza Méndez and Pablo César Otero Acosta in database processing.

References

Aaltola, E. (2005) Animal ethics and interest conflicts. *Ethics & the Environment* 10(1), 19–48. DOI: 10.2979/ETE.2005.10.1.19.

Acharya, K.P. and Wilson, R.T. (2020) The animal sacrifice–public health nexus in Nepal. *Journal of Public Health Policy* 41(3), 386–389. DOI: 10.1057/s41271-020-00224-3.

Adams, C.J. (2020) *The Pornography of Meat*, New and Updated Edition. Bloomsbury Academic, London. DOI: 10.5040/9781501364433.

Äijälä, M. (2019) Knowing through interspecies relationality in tourism? Animal agency in human-sled dog encounters. *Finnish Journal of Tourism Research* 15(2), 45–50. DOI: 10.33351/mt.88270.

Alders, R.G., Campbell, A., Costa, R., Guèye, E.F., Ahasanul Hoque, M. *et al.* (2021) Livestock across the world: diverse animal species with complex roles in human societies and ecosystem services. *Animal Frontiers* 11(5), 20–29. DOI: 10.1093/af/vfab047.

Anzoátegui, M. (2022) Contra la trampa del antropocentrismo: su crítica contemporánea como construcción de sentido, sesgo epistemológico y supuesto de inevitabilidad (Against the trap or anthropocentrism: a contemporary criticism of its construction of meaning, epistemological bias, and inevitability assumptions). *Revista Latinoamericana de Estudios Críticos Animales* 1, 433–481.

Ávila, L. (2014) *El animal de la zootecnia. Políticas de la des/ontologización* (The animal of zootechnics. Policies of de/ontologization). Master's thesis, University of the Andes, Colombia.

Balmond, L., Regad, C. and Riot, C. (2019) *The Toulon Declaration. The Legal Personality of Animals* (II). Faculty of Law, University of Toulon, Toulon, France.

Bostrom, N. (2005) A history of transhumanist thought. *Journal of Evolution and Technology* 14(1), 1–25. DOI: 10.5840/jpr_2005_26.

Büscher, B. (2018) From biopower to ontopower? Violent responses to wildlife crime and the new geographies of conservation. *Conservation and Society* 16(2), 157–169. DOI: 10.4103/cs.cs_16_159.
Çalik, A.Ö. and Çiftçi, G. (2013) Animal ethics in tourism. *International Journal of Business and Management Studies* 5(1), 43–53.
Cancino, Á. and De Alba, M. (2018) La ecoética como respuesta a las crisis antropogénicas (Ecoethics as a response to anthropogenic crises). In: *Zooética: Una Mirada Filosófica a Los Animales*. FCE, Mexico City, Mexico, pp. 102–137.
Carr, N. and Broom, D.M. (eds) (2018) *Tourism and Animal Welfare*. CAB International, Wallingford, UK. DOI: 10.1079/9781786391858.0000.
Carr, N. and Cohen, S. (2009) Holidaying with the family pet: no dogs allowed! *Tourism and Hospitality Research* 9(4), 290–304. DOI: 10.1057/thr.2009.10.
Castree, N. and Nash, C. (2006) Posthuman geographies. *Social & Cultural Geography* 7(4), 501–504. DOI: 10.1080/14649360600825620.
Chavarría, G. (2015) El posthumanismo y los cambios en la identidad humana (Posthumanism and changes in human identity). *Revista Reflexiones* 94(1), 97–107. DOI: 10.15517/rr.v94i1.20882.
Ćirović, D., Penezić, A. and Krofel, M. (2016) Jackals as cleaners: ecosystem services provided by a mesocarnivore in human-dominated landscapes. *Biological Conservation* 199, 51–55. DOI: 10.1016/j.biocon.2016.04.027.
Colella, S. (2018) Not a mere tangential outbreak: gender, feminism and cultural heritage. *Il Capitale Culturale Studies on the Value of Cultural Heritage* 18, 251–275. DOI: 10.13138/2039-2362/1897.
Cole-Turner, R. (2022) Posthumanism and transhumanism. In: *Encyclopaedia of Religious Ethics*. Wiley Online Library, Chichester, UK, pp. 1098–1105.
Cottingham, J. (1978) 'A brute to the Brutes?': Descartes' treatment of animals. *Philosophy* 53(206), 551–559. DOI: 10.1017/S0031819100026371.
Cudworth, E., Hobden, S. and Kavalski, E. (2018) *Posthuman Dialogues in International Relations*. Routledge, Abingdon, UK. DOI: 10.4324/9781315613475.
De Garis, H. (2005) *The Artilect War: Cosmists vs. Terrans. A Bitter Controversy Concerning Whether Humanity Should Build Godlike Massively Intelligent Machines*. ETC Books, New York.
Diedrich, A. (2007) The impacts of tourism on coral reef conservation awareness and support in coastal communities in Belize. *Coral Reefs* 26(4), 985–996. DOI: 10.1007/s00338-007-0224-z.
Di Pietro, L., Guglielmetti Mugion, R. and Renzi, M.F. (2018) Heritage and identity: technology, values and visitor experiences. *Journal of Heritage Tourism* 13(2), 97–103. DOI: 10.1080/1743873X.2017.1384478.
Doiron, S. and Weissenberger, S. (2014) Sustainable dive tourism: social and environmental impacts – The case of Roatan, Honduras. *Tourism Management Perspectives* 10, 19–26. DOI: 10.1016/j.tmp.2013.12.003.
Duffy, R. and Moore, L. (2011) Global regulations and local practices: the politics and governance of animal welfare in elephant tourism. *Journal of Sustainable Tourism* 19(4–5), 589–604. DOI: 10.1080/09669582.2011.566927.
Dumoulin, D. (2002) Nuevas imbricaciones entre patrimonio cultural y patrimonio natural. *TRACE (Travaux et Recherches Dans Les Amériques Du Centre)* 40, 59–64. DOI: 10.22134/trace.40.2001.546.
Elias, N. (2000) *Civilizing Process: Sociogenetic and Psychogenetic Investigations*. Blackwell Publishing, Oxford, UK.
FAADA (2023) Patrimonio animal y turismo (Animal heritage and tourism). Foundation for Advice and Action in the Defence of Animals. Available at: https://faada.org/docs/Informe_PATRIMONIOANIMALYTURISMO.pdf (accessed 16 May 2023).
Faria, C. and Paez, E. (2014) Anthropocentrism and speciesism: conceptual and normative issues. *Revista Bioetica & Derecho* 32, 95–103. DOI: 10.4321/S1886-58872014000300009.
Faria, C. (2016) Lo personal es político: feminismo y antiespecismo (The personal is political: feminism and antispeciesism). *Revista Latinoamericana de Estudios Críticos Animales* 3, 20–38.
Fennell, D. (2012a) *Tourism and Animal Ethics*. Routledge, Oxford, UK. DOI: 10.4324/9780203153642.
Fennell, D. (2012b) Tourism, animals and utilitarianism. *Tourism Recreation Research* 37(3), 239–249. DOI: 10.1080/02508281.2012.11081712.
Fennell, D. (2015) Akrasia and tourism: why we sometimes act against our better judgement? *Tourism Recreation Research* 40(1), 95–106. DOI: 10.1080/02508281.2015.1006419.
Ferrando, F. (2013) Posthumanism, transhumanism, antihumanism, metahumanism, and new materialisms. *Existenz* 8(2), 26–32.

Ferrando, F. (2019) *Philosophical Posthumanism*. Bloomsbury, London, UK, New York. DOI: 10.5040/9781350059511.

Fleisner, P. (2019) El animal como medio. Notas sobre zoopolíticas artísticas. *Tabula Rasa* 31, 77–97. DOI: 10.25058/20112742.n31.03.

Fletcher-Barnes, E. (2020) Cuddle, kill, conserve: a posthuman analysis of the African lion within the South African wildlife security assemblage. *International Journal of Sociology and Social Policy* 41(3–4), 475–488. DOI: 10.1108/IJSSP-09-2019-0187.

Francione, G. (1995) *Animals, Property, and the Law*. Temple University Press, Philadelphia, Pennsylvania.

Francione, G. (2000) *Introduction to Animal Rights: Your Child or the Dog?* Temple University Press, Philadelphia, Pennsylvania.

Frost, W. (2011) *Zoos and Tourism: Conservation, Education, Entertainment*. Channel View Publications, Bristol, UK. DOI: 10.21832/9781845411657.

Gallego, A.M. (2018) El arquetipo del guerrero como factor determinante para la creación de la masculinidad hegemónica. *Revista Indisciplinas* 4(7), 59–78.

Gâştescu, P. and Toma, E. (2019) Danube delta biosphere reserve. Tourist potential, turning to good account, impact. *Riscuri si Catastrofe* 25(2), 7–32. DOI: 10.24193/RCJ2019_11.

Giglio, V.J., Luiz, O.J. and Ferreira, C.E. (2020) Ecological impacts and management strategies for recreational diving: a review. *Journal of Environmental Management* 256, 109949. DOI: 10.1016/j.jenvman.2019.109949.

Graham, B., Ashworth, G.J. and Tunbridge, J.E. (2000). *A Geography of Heritage: Power, Culture and Economy*. Routledge, New York.

Green, R. and Giese, M. (2004) Negative effects of wildlife tourism on wildlife. In: Higginbottom, K. (ed.) *Wildlife Tourism: Impacts, Management and Planning*. Common Ground Publishing Party Ltd, Victoria, Australia, pp. 81–97.

Greffe, X. (2011) L'économie politique du patrimoine culturel: de la médaille au rhizome. In: *ICOMOS 17th General Assembly, 2011-11-27 / 2011-12-02*, Paris, France. Available at: http://openarchive.icomos.org/id/eprint/1307/1/IV-3-Article6_Greffe.pdf (accessed 14 May 2023).

Horta, O. (2007) *Un Desafío para la Bioética. La Cuestión del Especismo*. Universidad de Santiago de Compostela, Galicia, Spain.

Horta, O. (2010) What is speciesism? *Journal of Agricultural and Environmental Ethics* 23(3), 243–266. DOI: 10.1007/s10806-009-9205-2.

Horta, O. (2011) La argumentación de Singer en liberación animal: concepciones normativas, interés en vivir y agregacionismo. *Diánoia* 56(67), 65–85. DOI: 10.22201/iifs.18704913e.2011.67.171.

Horta, O. (2017) Why the concept of moral status should be abandoned. *Ethical Theory and Moral Practice* 20(4), 899–910. DOI: 10.1007/s10677-017-9829-7.

Hosey, G., Melfi, V. and Ward, S.J. (2020) Problematic animals in the zoo: the issue of charismatic megafauna. In: Angelici, F. and Rossi, L. (eds) *Problematic Wildlife II: New Conservation and Management Challenges in the Human-Wildlife Interactions*. Spinger, Cham, Switzerland, pp. 485–508. DOI: 10.1007/978-3-030-42335-3.

Hribal, J. (2003) "Animals are part of the working class": a challenge to labor history. *Labor History* 44(4), 435–453. DOI: 10.1080/0023656032000170069.

Jiménez, G. (2016) Des-patriarcalizar el patri-monio: una propuesta desde el feminismo. *La Descomunal. Revista Iberoamericana de Patrimonio y Comunidad* 2(2), 338–345.

Joy, M. (2010) *Why We Love Dogs, Eat Pigs, and Wear Cows: an Introduction to Carnism*. Red Wheel, Newburyport, Massachusetts.

Kaltofen, C. (2018) With a posthuman touch: international relations in dialogue with the posthuman – a human account. In: Cudworth, E., Hobden, S. and Kavalski, E. (eds) *Posthuman Dialogues in International Relations*. Routledge, London and New York, pp. 35–51. DOI: 10.4324/9781315613475.

Kline, C. (ed.) (2018) *Animals, Food, and Tourism*. Routledge, Abingdon, UK. DOI: 10.4324/9781315265209.

Kline, C. (2023) Centering animals within rural tourism. In: Mair, H. (ed.) *Handbook on Tourism and Rural Community Development*. Edward Elgar Publishing, Cheltenham, UK, pp. 344–358. DOI: 10.4337/9781800370067.

Leggat, P.A. and Speare, R. (2000) Traveling with pets. *Journal of Travel Medicine* 7(6), 325–329. DOI: 10.2310/7060.2000.00087.

Leiper, N. (1979) The framework of tourism. *Annals of Tourism Research* 6(4), 390–407. DOI: 10.1016/0160-7383(79)90003-3.

Levy, J.E. (2013) Gender, feminism, and heritage. In: Biehl, P. and Prescott, C. (eds) *Heritage in the Context of Globalization*. Springer, New York, pp. 85–91. DOI: 10.1007/978-1-4614-6077-0.

Littlefield, J. (2010) Men on the hunt: ecofeminist insights into masculinity. *Marketing Theory* 10(1), 97–117. DOI: 10.1177/1470593109355250.

López, Á. and Quintero, G.J. (2021a) La geografía del turismo y la geografía de los animales intersectadas por la ética poshumanista (The geography of tourism and the geography of animals intersected by posthumanist ethics). *Cuadernos de Geografía: Revista Colombiana de Geografía* 30(1), 86–105. DOI: 10.15446/rcdg.v30n1.82631.

López, Á. and Quintero, G.J. (2021b) Animal dark tourism in Mexico: bulls performing their own slaughter. In: Rickly, J. and Kline, C. (eds) *Exploring Non-Human Work in Tourism: From Beasts of Burden to Animal Ambassadors*. De Gruyter, Berlin, pp. 69–82. DOI: 10.1515/9783110664058.

Low, P. (2012) The Cambridge Declaration on Consciousness. Available at: http://fcmconference.org/img/CambridgeDeclarationOnConsciousness.pdf (accessed 18 May 2023).

Lowenthal, D. (1998) *The Heritage Crusade and the Spoils of History*. Cambridge University Press, Cambridge, UK. DOI: 10.1017/CBO9780511523809.

Luke, B. (1998) Violent love: hunting, heterosexuality, and the erotics of men's predation. *Feminist Studies* 24(3), 627–655. DOI: 10.2307/3178583.

Lynn, W. (1998) Animals, ethics and geography. In: Wolch, J. and Emel, J. (eds) *Animal Geographies: Place, Politics and Identity in the Nature-Culture Borderlands*. Verso, London, pp. 280–298.

Maciejewski, K. and Kerley, G.I. (2014) Understanding tourists' preference for mammal species in private protected areas: is there a case for extralimital species for ecotourism? *PloS One* 9(2), e88192. DOI: 10.1371/journal.pone.0088192.

Markwell, K. (ed.) (2015) *Animals and Tourism: Understanding Diverse Relationships*. Channel View Publications, Bristol, UK. DOI: 10.21832/9781845415051.

Massey, J.E. and Randriamarohaja, R.H. (2019) Blackfish: a crisis of legitimacy at seaworld. *Tourism Culture & Communication* 19(3), 195–207. DOI: 10.3727/194341419X15554157596100.

McLennan, C.J., Becken, S. and Moyle, B.D. (2017) Framing in a contested space: media reporting on tourism and mining in Australia. *Current Issues in Tourism* 20(9), 960–980. DOI: 10.1080/13683500.2014.946893.

Miller, M.R. (2013) Descartes on animals revisited. *Journal of Philosophical Research* 38, 89–114. DOI: 10.5840/jpr2013386.

Moorhouse, T., D'Cruze, N.C. and Macdonald, D.W. (2017) Unethical use of wildlife in tourism: what's the problem, who is responsible, and what can be done? *Journal of Sustainable Tourism* 25(4), 505–516. DOI: 10.1080/09669582.2016.1223087.

Morris, M. (2019) Injured roos increasingly left to die on country roads, wildlife carer says, as drivers 'don't know what to do'. ABC News. Available at: https://www.abc.net.au/news/2019-05-26/concern-injured-kangaroos-left-to-suffer-on-country- roads/11131260 (accessed 28 May 2023).

Muzur, A. and Rinčić, I. (2015) Two kinds of globality: a comparison of Fritz Jahr and Van Rensselaer Potter's bioethics. *Global Bioethics* 26(1), 23–27. DOI: 10.1080/11287462.2015.1007616.

Nielsen, K. (2008) The private parts of animals: Aristotle on the teleology of sexual difference. *Phronesis* 53(4–5), 373–405. DOI: 10.1163/156852808X338337.

Nilsson, P.A. (2018) Impact of cultural heritage on tourists. The heritagization process. *Athens Journal of Tourism* 5(1), 35–54. DOI: 10.30958/ajt.5.1.3.

Nurse, A. (2020) Masculinities and animal harm. *Men and Masculinities* 23(5), 908–926. DOI: 10.1177/1097184X20965458.

Oliver, C. (2022) *Veganism, Archives, and Animals: Geographies of a Multispecies World for Social and Cultural Geography*. Routledge, London. DOI: 10.4324/9781003141211.

Orlić, I. and Marjanić, S. (2018) Heritage animals – why not? Anthropocentrism notwithstanding. *Studia Mythologica Slavica* 21, 189–215. DOI: 10.3986/sms.v21i0.7073.

Ormrod, J.S. (2011) 'Making room for the tigers and the polar bears': biography, phantasy and ideology in the Voluntary Human Extinction Movement. *Psychoanalysis, Culture & Society* 16(2), 142–161. DOI: 10.1057/pcs.2009.30.

Panko, T.R. and George, B.P. (2018) Animal sexual abuse and the darkness of touristic immorality. In: Korstanje, M. and George, B. (eds) *Virtual Traumascapes and Exploring the Roots of Dark Tourism*. IGI Global, Hershey, Pennsylvania, pp. 175–189.

Perez, G. (2020) Materiales para defender los derechos de los animales liminales (Materials to defend liminal animal rights). *Revista Latinoamericana de Estudios Críticos Animales* 7(2), 21–84.

Quintero, G.J. and López, A. (2018) Tauromaquia y turismo oscuro en México: las corridas de toros como prácticas no éticas. *Teoría y Praxis* 24, 197–227.

Quintero, G.J. and López, Á. (2020) Geografía de los animales: construcción filosófica de una subdisciplina científica a través de su historia. *Cuadernos de Geografía: Revista Colombiana de Geografía* 29(1), 16–31. DOI: 10.15446/rcdg.v29n1.78653.

Rakić, T. (2011) Philosophies of the visual [method] in tourism research. In: Rakić, T. and Chambers, D. (eds) *An Introduction to Visual Research Methods in Tourism*. Routledge, Oxford, UK, pp. 17–32. DOI: 10.4324/9780203855867.

Regan, T. (1983) *The Case for Animal Rights*. University of California Press, Berkeley, California.

Reynolds, P.C. and Braithwaite, D. (2001) Towards a conceptual framework for wildlife tourism. *Tourism Management* 22(1), 31–42. DOI: 10.1016/S0261-5177(00)00018-2.

Rickly, J.M. and Kline, C. (eds) (2021) *Exploring Non-human Work in Tourism: From Beasts of Burden to Animal Ambassadors*. De Gruyter, Berlin.

Roigé, X. and Frigolé, J. (2010) *Constructing Cultural and Natural Heritage: Parks, Museums and Rural Heritage*. ICRPC, Girona, Spain.

Ryder, R.D. (1975) *Victims of Science: The Use of Animals in Research*. Davis-Poynter, London.

Salmen, A. and Dhont, K. (2023) Animalizing women and feminizing (vegan) men: the psychological intersections of sexism, speciesism, meat, and masculinity. *Social and Personality Psychology Compass* 17(2), e12717. DOI: 10.1111/spc3.12717.

Sass, H. (2007) Fritz Jahr's 1927 concept of bioethics. *Kennedy Institute of Ethics Journal* 17(4), 279–295. DOI: 10.1353/ken.2008.0006.

Shani, A. and Pizam, A. (2009) Tourists' attitudes toward the use of animals in tourist attractions. *Tourism Analysis* 14(1), 85–101. DOI: 10.3727/108354209788970225.

Singer, P. (1975) *Animal Liberation*, Random House, New York.

Singer, P. (1987) Animal liberation or animal rights? *The Monist* 70(1), 3–14. DOI: 10.5840/monist19877018.

Singer, P. (2009) Speciesism and moral status. *Metaphilosophy* 40(3–4), 567–581. DOI: 10.1111/j.1467-9973.2009.01608.x.

Skibins, J.C., Powell, R.B. and Hallo, J.C. (2013) Charisma and conservation: charismatic megafauna's influence on safari and zoo tourists' pro-conservation behaviors. *Biodiversity and Conservation* 22(4), 959–982. DOI: 10.1007/s10531-013-0462-z.

Skibins, J.C., Dunstan, E. and Pahlow, K. (2017) Exploring the influence of charismatic characteristics on flagship outcomes in zoo visitors. *Human Dimensions of Wildlife* 22(2), 157–171. DOI: 10.1080/10871209.2016.1276233.

Srinivasan, K. and Kasturirangan, R. (2017) Conservation and invasive alien species: violent love. In: Maher, J., Pierpoint, H. and Beirne, P. (eds) *The Palgrave International Handbook of Animal Abuse Studies*. Palgrave Macmillan, London, pp. 433–452.

Steffen, W., Grinevald, J., Crutzen, P. and McNeill, J. (2011) The anthropocene: conceptual and historical perspectives. *Philosophical Transactions of the Royal Society A: Mathematical, Physical, and Engineering Sciences* 369(1938), 842–867. DOI: 10.1098/rsta.2010.0327.

Sterling, C. (2020) Critical heritage and the posthumanities: problems and prospects. *International Journal of Heritage Studies* 26(11), 1029–1046. DOI: 10.1080/13527258.2020.1715464.

Taş Gürsoy, İ. (2020) Horse-drawn carriages: sustainability at the nexus of human-animal interaction. *Journal of Sustainable Tourism* 28(2), 204–221. DOI: 10.1080/09669582.2019.1671852.

Taylor, N. and Signal, T.D. (2009) Pet, pest, profit: isolating differences in attitudes towards the treatment of animals. *Anthrozoös* 22(2), 129–135. DOI: 10.2752/175303709X434158.

Taylor, P. (2011) *Respect for Nature: a Theory of Environmental Ethics*, Princeton University Press, Princeton, New Jersey.

Tickle, L. and von Essen, E. (2020) The seven sins of hunting tourism. *Annals of Tourism Research* 84, 102996. DOI: 10.1016/j.annals.2020.102996.

Tucker, H. and Shelton, E. (2014) Traveling through the end times: the tourist as apocalyptic subject. *Tourism Analysis* 19(5), 645–654. DOI: 10.3727/108354214X14116690098133.

Tunbridge, J.E., Ashworth, G.J. and Graham, B.J. (2013) Decennial reflections on a geography of heritage (2000). *International Journal of Heritage Studies* 19(4), 365–372. DOI: 10.1080/13527258.2012.695038.

Twine, R. (2021) Masculinity, nature, ecofeminism, and the "Anthropo"cene. In: Pulé, P.M. and Hultman, M. (eds) *Men, Masculinities, and Earth: Contending With the (m) Anthropocene*. Palgrave Macmillan, London, pp. 117–133. DOI: 10.1007/978-3-030-54486-7.

UNESCO (2023a) The world heritage convention. Available at: https://whc.unesco.org/en/convention/ (accessed 28 April 2023).

UNESCO (2023b) Intangible cultural heritage. Available at: https://ich.unesco.org/en (accessed 28 April 2023).

Urbanik, J. and Johnston, C.L. (2017) *Humans and Animals: A Geography of Coexistence*. ABC-CLIO, Santa Barbara, California. DOI: 10.5040/9798400667565.

Vint, S. (2010) Animal studies in the era of biopower. *Science Fiction Studies* 37(3), 444–455.

Weaver, J.A. (2009) The posthuman condition: a complicated conversation. In: Malewski, E. (ed.) *Curriculum Studies Handbook – The Next Moment*. Routledge, New York, pp. 208–224. DOI: 10.4324/9780203877791.

Weitzenfeld, A. and Joy, M. (2014) An overview of anthropocentrism, humanism, and speciesism in critical animal theory. *Counterpoints* 448, 3–27.

Whiting, M.J., Williams, V.L. and Hibbitts, T.J. (2013) Animals traded for traditional medicine at the Faraday market in South Africa: species diversity and conservation implications. In: Alves, R. and Rosa, I. (eds) *Animals in Traditional Folk Medicine Implications for Conservation*. Springer, Berlin, pp. 421–473.

Winter, C. (2020) A review of research into animal ethics in tourism: launching the annals of tourism research curated collection on animal ethics in tourism. *Annals of Tourism Research* 84, 102989. DOI: 10.1016/j.annals.2020.102989.

Wolfe, C. (2010) *What is Posthumanism?* University of Minnesota Press, Minneapolis, Minnesota.

3 The Legacy of Zoos? Moral Reasons against Considering Zoos as Heritage

Gustavo Ortiz-Millán*
Universidad Nacional Autónoma de México, Mexico City, Mexico

Abstract
Some people have held that zoos should be seen as heritage tourist attractions. However, some others wonder whether they are ethically justifiable. In this text I present the main arguments on which the discussion around zoos is based. Five of the arguments used to defend the existence of zoos are: (i) they are places of recreation; (ii) they are heritage tourism attractions; (iii) zoos fulfil an educational function; (iv) zoos do research; and (v) zoos do conservation work, that is, they protect and reintroduce wildlife species at risk of extinction. I review some of the central arguments both for and against each of these points. After taking stock of these arguments, we can conclude that there are strong reasons to take a sceptical position against the arguments that justify the existence of zoos. Finally, zoos were created, and still are justified, with a set of mind that sees animals as commodities, as mere resources we can use for our own purposes. I conclude that if we recognize that on the same basis on which we attribute rights to humans we can attribute rights to animals, then the answer to the question that motivates this chapter is that zoos are not ethically justifiable.

Introduction

Zoos have been the subject of intense debate as to whether they are ethically justifiable. To take a stand on this issue and be able to ethically evaluate their existence, it is necessary to know and analyse critically the arguments that are given for and against them. In this text I will present the main historical arguments on which the discussion around zoos is based. We are told that they are valuable because: (i) they are places of recreation and tourism, some zoos even being designated heritage tourist attractions (see the 2008 special issue of the *International Journal of Heritage Studies* devoted to zoos as heritage); (ii) they fulfil an educational function (Godinez and Fernandez, 2019); (iii) they do research (Association of Zoos and Aquariums, 2023a); and (iv) they do conservation work, that is, they protect and reintroduce wildlife species at risk of extinction (Association of Zoos and Aquariums, 2023b). Those who oppose zoos have criticized these arguments (Jamieson, 2004a, b; Margodt, 2010). I will briefly review some of the central arguments both for and against each of these points. The premise of this chapter therefore is to critically assess the 'heritage' of zoos and question whether this is an institution that must be preserved. After taking stock of these arguments, it seems that there are strong reasons to take a sceptical position against the arguments that justify the existence of zoos. Many may see zoos as part of our inherited traditions and institutions, but I shall argue that there are strong moral reasons to get rid of this cultural heritage.

*gmom@filosoficas.unam.mx

 Tourism, Heritage and Commodification of Non-human Animals (eds A. López López *et al.*)
DOI: 10.1079/9781800623309.0003

I do not only want to show the weakness of the arguments in favour of zoos. My approach is based on the idea of animal rights, and from that perspective I shall argue that zoos constitute a violation of the rights of animals. My approach can be called posthumanist in the sense that it argues that on the same basis on which we attribute rights to humans, we can attribute rights to animals. I argue that animals have fundamental rights to life and liberty. On the other hand, we humans have no inherent right to deprive animals of their liberty by confining them in a zoo or to dispose of their lives if it is in our best interest to do so – particularly when the reasons for keeping them in captivity are not as strong as zoo advocates want to show. Human rights coexist on a spectrum with animal rights (Evans, 2015), and as it happens with rights in general, human and animal rights tend to collide with each other. There is no reason to think that human rights must always take precedence over those of non-human animals. These must be balanced in each case. However, when human interests tend to be questionable, as in the case of zoos, there are strong reasons for thinking that animal rights must take precedence.

A Brief History of Zoos

Zoos are institutions that exhibit animals in captivity to the public for recreational, touristic, educational, conservation and research purposes. Sometimes zoos and their associated facilities also breed and raise animals. If we understand them in this way, then zoos are a relatively recent institution in history. Since antiquity, there have been private collections of animals, but they had no educational, recreational or touristic functions, they were just a symbol of power and wealth. They were not zoos, properly speaking, as they are currently defined. The first zoos as such appeared in the 18th century, when private collections of beasts (*ménageries*) gave way to zoological gardens, such as those of Schönbrunn in Vienna in 1765, the Jardin des Plantes in Paris in 1793 and the London Zoological Garden in 1828, which was the first one to call itself 'zoo' (Kisling, 2001). The appearance of zoos in the 19th century was also due to the fact that they brought considerable prestige to the cities that housed them, typically capitals of imperial metropolises. As in antiquity, they were symbols of power and wealth. As John Berger explains in his classic essay 'Why look at animals?':

> In the 19th century, public zoos were an endorsement of modern colonial power. The capturing of the animals was a symbolic representation of the conquest of all distant and exotic lands. 'Explorers' proved their patriotism by sending home a tiger or an elephant. The gift of an exotic animal to the metropolitan zoo became a token in subservient diplomatic relations. Yet, like every other 19th century public institution, the zoo, however supportive of the ideology of imperialism, had to claim an independent and civic function.
>
> (Berger, 1992, p. 19)

The function was to expand the scientific knowledge and education of the people. On the other hand, just as our modern idea of the zoo was born with colonialist imperialism, coincidentally, our modern idea of tourism was born with it, in which zoos began to have an increasingly important place, and at some point they were seen as part of societies' cultural heritage.

In the USA, the first European-style zoo was founded in Philadelphia in 1874. In Latin America, the first zoos as such were founded in Buenos Aires (1875), Rio de Janeiro (1888) and Montevideo (1894). Later, the Chapultepec Zoo was founded in Mexico City in 1923 and the National Zoo of Chile in Santiago in 1925. Nowadays, there are some 10,000 zoos worldwide, with around one million vertebrate animals, which receive some 700 million visitors each year (Gusset and Dick, 2011).

Although I analyse zoos here, most of what is said can also be extended to aquariums. While they have specific issues depending on the type of animals they handle, there are no significant differences between the arguments in favour of and against either.

Recreation, Tourism and Education

The public who attend zoos identify them as places of recreation and tourism, therefore a strong component of the heritage of zoos is that of being a place to visit when travelling.

Traditionally, people have taken their families to the zoo over the weekend, on vacations or on days off to spend the day. The vast majority of visitors see their experience at the zoo as a social activity with strong benefits of family cohesion. Two-thirds of visitors to zoos are families and the main reason to visit them is to make a 'family outing' (Cain and Meritt, 2007). Many zoos advertise themselves as places of recreation and tourism, given that in many cases, whether public or private institutions, they depend on the income thus generated. Many even explicitly offer the organization of birthday parties, weddings and congresses, as well as opportunities to interact with animals, ride them or feed them, among many other activities (Carr and Cohen, 2011). Several cities have encouraged, as part of their tourism policies, the creation of zoos, which have become one of the main touristic attractions, especially for families with children (Frost, 2011). In some cases, between 60 and 75% of visitors are people who do not live in the city where the zoo is located and who are doing tourism (Cain and Meritt, 2007). In fact, tourism is one of the main reasons for promoting the creation of zoos – and not so much the promotion of education, research or conservation, which will later be cited as their main justifying reasons. Linke and Winter summarize the idea of zoos as tourist attractions as follows:

> Zoos also contribute significantly to the national economy. Operating as a tourist attraction, they create employment, purchase goods and services, receive overseas tourists which instigates foreign exchange, reinvest any surpluses in zoo development, and also directly and indirectly attract visitors to the local area.
>
> (Linke and Winter, 2011, pp. 69–70; for a broader picture of the relationship between zoos and tourism, see Tribe, 2004; Frost, 2011)

Some cities have aimed at marketing their zoos as heritage tourist attractions, especially those with historical zoos, and as places that preserve the natural heritage of the country/region, describing themselves as modern Noah's arks, as sanctuaries for endangered species (Mason, 1999; *International Journal of Heritage Studies*, 2008). However, the fact that zoos are touristic attractions cannot serve as a basis for their ethical justification, because there are many other ways to attract tourists to a city that do not involve the captivity of animals (such as the creation of museums, theme parks, festivals, monuments).

Some studies show that many people who take the family to the zoo for 'a day out' are not particularly interested in animals. A sociological study, conducted by Edward Ludwig, concluded that 'for most people, most animals are not interesting enough to command more than fleeting attention. Indeed, most animals are likely to be viewed in passing unless the animal does something to bring the spectator to a halt' (1981, p. 314). People tend to stop when animals are fed, make noise, imitate human behaviour or when there are offspring. Otherwise, according to Kreger (1998, p. 378), people stop between 30 s and 2 minutes at each exhibit, which is not enough time to read the signs. In a different study on the time people spent at the National Zoo in the USA, Marcellini and Jenssen (1988) tell us that visitors spent less than 8 seconds per snake and about 2 seconds with amphibians. People's indifference to animals, says Ludwig, also tends to discourage zoo employees. Other research found that only 27% of people bothered to read the signs and interpretive material at exhibits (Clayton *et al.*, 2008). Some studies on visitors show that a low percentage of visitors are interested in learning about animals or conservation (Dunlap and Kellert, 2004).

Although recreation and tourism are often cited as part of the justification for the existence of zoos, since the 1970s the attitude of people towards this institution has changed and there has been greater concern for the welfare of animals in captivity, which is why zoos have been forced to present themselves as educational, research and conservation institutions to justify their existence. Neither recreation nor tourism are strong enough arguments to keep thousands of animals captive, often in questionable welfare conditions, especially because there are currently many other forms of recreation that do not have the problems of moral justification that zoos have.

Many emphasize the educational character of zoos. It is argued that since more people live in cities, the connection with wild animals has been lost and zoos offer a good opportunity for people to know animals directly and learn about nature. In addition, given the great destruction of the habitat of many species around the world, zoos offer one of the few opportunities to see some animals whose habitats no longer exist. For this reason, zoos emphasize their educational aspect (Patrick *et al.*, 2007) and in fact have redirected their efforts to encourage not only children but also people of any age to learn about animals. Likewise, they have increased their pedagogical tools so that now they use more information technologies, videos and interactive screens that allow the visitor to have a richer experience (Andersen, 2003). However, few zoos around the world have this type of educational tools and most of the information received by visitors still comes from the explanatory signs placed next to the exhibits – which most people do not read. Many zoos also do not have enough staff to do guided tours, even though these visits improve learning in the visitors. (Eric Jensen evaluated the results, among children aged 7 to 15 years, related to learning about biology and conservation from guided tours and tours not guided by an educator at the London Zoo. He stated: '41% of educator guided visits and 34% of unguided visits resulted in conservation biology-related learning. Negative changes in children's understanding of animals and their habitats were more prevalent in unguided zoo visits' (2014, p. 1004). In fact, these numbers suggest that in both cases the educational impact is low.)

However, despite the great emphasis that zoos put on education and the growing literature on their educational role, there is little evidence to show that visits to the zoo or even direct interaction with animals improve visitors' knowledge about biology, or issues related to conservation (Clayton *et al.*, 2008). There is also no evidence that visits to zoos have a greater educational impact on visitors than other means of information about animals, for example, television educational programmes about wild animals. If someone can learn more about an animal species or about biodiversity through a National Geographic or an Animal Planet documentary than a visit to the zoo, then the weight of the educational argument in favour of zoos loses strength.

There are alternative ways of education about animals: visits to natural history museums or web pages with data about wild animals, among others. However, those who defend zoos argue that, although all this may be true, this type of interaction is 'impersonal, vicarious and voyeuristic' (Patrick and Tunnicliffe, 2013, p. 179) (although one might wonder whether a visit to the zoo is not equally voyeuristic). They are two-dimensional images, they say, reduced in size, which do not provoke emotional, affective or aesthetic reactions at first hand. However, there are forms of knowledge about animals that do involve first-hand knowledge, such as field visits aimed at knowing their habitat in ways that are non-invasive or not harmful to animals, as is the case with non-invasive ecotourism. Unfortunately, we lack studies showing that one can learn more through these other options than from a visit to the zoo, but that is an investigation that zoos must do, that is, they must show that people learn more through them than from other means. As Moss and Esson say, the burden of evidence on the educational role of zoos falls on their shoulders:

> Zoos and their accreditation bodies find themselves between a rock and a hard place. For many years, they have confidently promoted themselves as education providers particularly with regard to the conservation of biodiversity; perhaps even used this educational function as part justification for their existence. Because of this, the burden of evidencing educational impact falls squarely on the shoulders of zoos. Yet the research undertaken thus far (and there is a substantial amount) has clearly not been universally accepted as an effective demonstration of zoos' positive impact. Indeed, the peer-reviewed criticisms of recent years ... would suggest that this issue is here to stay. Away from strategic educational goals and mission statements, we found evidence to suggest that accreditation bodies and some zoos were making public, causal claims about zoos, and their positive educational impact. The larger the claims, the larger the evidence-base required to support them.
>
> (Moss and Esson, 2013, p. 17)

Zoos' educational character should not consist simply of giving information to visitors, but in promoting positive attitudes and values towards animals and nature, that is, in promoting a proper environmental ethic. In that sense, some critics of zoos have argued that, in fact, zoos have a negative educational impact. Many years ago, Robert Sommer argued that many zoos tend to perpetuate people's stereotypes of animals, especially children's, for example that wolves are ferocious or that hyenas are cowardly and contemptuous predators or that sloths are lazy (or other forms of anthropomorphizing that do not help to understand animal behaviour). 'Despite excellent intentions, even the best zoos may be creating animal stereotypes that are not only incorrect but that actually work against the interests of wildlife preservation' (Sommer, 1972, pp. 27–28). Although the situation has changed since Sommer wrote this, many zoos do not do much to change the stereotypical ways people tend to view animals. In any case, if there is an implicit message in zoos it is that it is correct to keep animals in captivity if that is justified with an excuse (for example, that it is done for educational or conservation purposes). They reinforce the idea that animals exist to serve human beings and not as conscious individuals with their own interests, and with a right to live in their natural environments. They perpetuate the idea that non-human animals are inferior because humans can control their lives and lock them up to serve our purposes. (In 2014 the World Association of Zoos and Aquariums (WAZA) published a global survey on the impact of zoos on the understanding of biodiversity. Although the report gives many positive data on how they help to better understand biodiversity, it also shows that visits to the zoo make people 17% less committed to adopt protection and habitat creation measures and 9% less likely to act against pollution and climate change (Parker, 2017, p. 46).)

Scientific Research

One of the most commonly cited reasons in favour of the existence of zoos is that they carry out scientific research. According to Anderson *et al.* (2008), zoos and aquariums around the world conduct research on different topics: 26.8% is on behaviour, 20.7% on reproduction, 11.3% on animal management, 8.3% on diet and nutrition, 7.3% on veterinary medicine, 5.9% on genetics, 1.8% on conservation and the rest on topics such as education, exhibit design, enrichment and population biology (see also Wharton, 2007; Loh *et al.*, 2018). However, not all zoos do research in all areas, and most of this research is done by a few wealthy zoos: Escribano *et al.* (2021) analysed the research articles published over a century by 447 institutions and found that 60% of research articles were authored by researchers from just 15 zoos. It is very likely that many zoos in low- or middle-income countries do little or no research at all.

Few zoos have full-time staff dedicated to research. While many industries devote 3 to 5% of their budget to research, zoos do not even reach that figure (Wharton, 2007, p. 182). The cost of maintaining a zoo is very high and dedicating money to research is a luxury that few can afford. Much of the research that is done in zoos is largely funded by other institutions, such as universities, private research institutions or government agencies, not by zoos. In those zoos that have personnel dedicated to research, many of them carry out research outside the zoo, for example, those who do research on conservation issues.

However, a more important issue than whether scientific research is done in zoos is the validity and utility of the information obtained. Some critics argue that the experimental design of most of the research conducted within zoos lacks scientific rigour and is rarely reported in peer-reviewed publications (Jamieson, 2004a). The problem is that very small populations are often studied and there is frequently a lack of a control group that is similar to the group studied. This means that the information obtained is not significant.

Much of the research that is done in zoos is done on ethology, that is, animal behaviour. However, it has been argued that there is not much to learn from wild animals in captivity and under unnatural conditions, for example, where they do not interact with other species, do not have to procure their own food or cannot develop many of their natural behaviours. It could be objected that ethological studies do provide knowledge about the behaviour of animals in captivity, for the benefit of these captive animals and that this knowledge is used to establish parameters to be able to compare these animals with their congeners in the wild. Obviously, if these animals were not captive, it would not be necessary to generate knowledge in the matter. Others have claimed that different topics can be investigated, such as domestication or possible behaviours that could not be manifested in freedom. It is possible that many zoos (especially those that have more space to reproduce more approximately the natural living conditions of animals) can generate more reliable data. However, it is more likely that much more accurate information can be generated by studying animals in their natural habitat, where they are in contact with their predators or their prey.

Other types of research (for example, anatomy or pathology) carried out in zoos have three main purposes: (i) to improve the conditions of zoos to optimize the conditions in which animals live and reproduce, (ii) to provide animal models that help us understand human and animal diseases and (iii) increase our knowledge about animals (Jamieson, 2004a, p. 170). However, although the first point sounds laudable, that would be unnecessary if wild animals were not kept in captivity in zoos. As Jamieson says, 'if there were no zoos, there would be no need to improve them' (2004a, p. 170).

On whether zoo animals can help us understand human diseases, we should remember that the bulk of the research done on animals is done with laboratory animals (such as mice, primates, cats or dogs) and not with wild animals such as those we find in a zoo. When wild animals are used, this research is not done in zoos but in laboratories. It is difficult to manipulate tigers, lions, giraffes or rhinos to carry out sometimes invasive procedures. Additionally, it is difficult to extrapolate the results of research carried out on an animal species to humans, given the many differences between species. In fact, there are few animals in which research can be done and whose results are useful for human beings.

The third purpose mentioned on research with zoo animals is to increase our knowledge about zoo animals. Undoubtedly, seeking knowledge is a valuable goal, but knowledge for its own sake is not valuable if it has no practical purpose. Those directly involved in the benefits of that knowledge are the animals themselves, since this knowledge would help us improve their health and well-being, but if millions of animals in captivity find their own welfare compromised precisely because they are in that situation, then there would be reasons to doubt the value of that knowledge.

Conservation

Conservation of endangered species and their habitats is for many the main function of zoos today. The Association of Zoos and Aquariums (AZA) defines zoos as conservation centres that are responsible for the health of the ecosystem and the survival of species. One of the requirements to belong to AZA is carrying out work on conservation. As of September 2022, there were 238 zoos throughout the world that were members of the association (Association of Zoos and Aquariums, 2023c). According to its Annual Report on Conservation and Science for 2020, AZA members invested US$208 million that year in conservation projects in 115 countries, and benefited 942 species or subspecies, 223 of which are in danger of extinction or threatened, listed under the U.S. Endangered Species Act (Association of Zoos and Aquariums, 2020). However, this is only a small amount (0.42%) compared to the US$4.9 billion they spend on operations and construction (Association of Zoos and Aquariums, 2019).

Many zoo advocates present them as modern Noah's arks conserving as many species as possible, especially in the face of the devastation of ecosystems around the world, global warming and the accelerated rate of species disappearance. Others present them as preserving the natural heritage of the region or the country. They usually cite examples of species that had disappeared in the wild and that were saved only thanks to their reintroduction from specimens that were kept in captivity, such as the Arabian oryx (*Oryx leucoryx*), the European bison (*Bison bonasus*), the black-footed ferret (*Mustela nigripes*) and the Mongolian wild horse (*Equus ferus przewalskii*) (Marris, 2021). Some zoos now have biobanks that preserve the genetic material of threatened species, in case these become extinct (Clarke, 2009). The survival plans of endangered species of AZA consider the protection of some 500 species (Association of Zoos and Aquariums, 2023d).

However, many critics of zoos doubt that their contribution to the conservation of species is significant. Successful reintroduction programmes, such as the ones already mentioned, are rare: they are very expensive, usually have little chance of success and require a viable natural habitat. One of the main causes of the extinction of species is the destruction of their habitats by anthropogenic causes: global warming, the destruction of forests to extend agriculture, but also excessive hunting and the negative impact of introduced species, among others. (This should lead us to ask if conservation of species makes sense if their natural habitats have been destroyed and these animals may never be reintroduced. However, this is a subject that requires a broader treatment than one I can give it here.) Zoos cannot protect more than an insignificant part of the thousands of species that are in danger of extinction. According to the Red List of Threatened Species of the IUCN (International Union for the Conservation of Nature), of the 150,388 species included, 42,100 (28%) are threatened with extinction (International Union for the Conservation of Nature, 2023). However, as mentioned, of these species, AZA currently has some 500 Species Survival Plans for threatened species.

There are few cases of animals in captivity that can be reintroduced into the wild: many only know captivity and are not prepared for the challenges of their release in natural habitats. For example, animals in captivity tend not to be afraid of humans because they have become accustomed to them and, when reintroduced, they tend to approach human populations in search of food or shelter, as has been seen, for example, with freed killer whales. Predators have to be taught how to hunt, because this behaviour is not instinctive but learned. In fact, it is very difficult, if not almost impossible, to release animals of many species, including endangered species such as elephants, polar bears, gorillas, orangutans, tigers and chimpanzees. In general, zoos do not breed animals with the intention of replacing threatened populations. Less than 30% of the species that zoos have are on the IUCN Red List as threatened species; in some cases, zoological collections do not even have 10% of threatened species (Zimmermann and Wilkinson, 2007, p. 307). In other words, most of the species that zoos have are not threatened (they are not vulnerable or in danger of extinction).

Within the percentage of threatened species that zoos have, most of their breeding programmes are focused on mammals (52.62%) and birds (35.62%). Reptiles, amphibians, fish and invertebrates occupy the remaining 12%. However, there does not seem to be any ecological reason for this distribution. According to the IUCN Red List data, the number of threatened amphibian species (41%) exceeds mammals (27%) or birds (13%) (International Union for the Conservation of Nature, 2023). Not only are amphibians more threatened, but the cost for their conservation is much less than that of any large mammal. The director of Amphibian Ark, Kevin Zippel, told *Scientific American* that amphibians are 'absolutely vital to their ecosystems' and that 'for the price to keep a single elephant in captivity for a year, about [US]$100,000, you could pay for the expertise and facilities to save an entire amphibian species' (cited by Margodt, 2010, p. 16). That is, zoos have a marked favouritism towards mammals and birds that are more charismatic, and probably attract more visitors, than towards amphibians or invertebrates, which usually are not so attractive to the public, even when their conservation would be more effective and less expensive. (Not only do they not encourage their conservation, but 59% of the reptiles and 73% of the amphibians mentioned in the 2003 ISIS (International Species Information System) database were captured from the wild, as were 21% of mammals and 37% of birds (Baker, 2007, p. 145). In fact, zoos occasionally need to take free-living animals for conservation purposes, e.g. to maintain the genetic diversity of a species.) The motivation to maintain and reproduce large mammals in zoos is not ecological, but has to do with the economic interests of zoos (many of them private institutions that depend on their income for survival), as well as what they perceive as visitor preferences and their aesthetic evaluations. This has led some zoo critics to argue that 'zoo conservation is in the first place about conserving zoos, not about conserving threatened species' (Margodt, 2010, p. 17, his emphasis).

The probabilities of success of the reintroduction programmes of species are very low. According to Beck *et al.* (1994), of 145 projects that released animals bred in captivity to re-establish or reinforce natural populations, only 16 (11%) were successful. In a more recent study, Bolam *et al.* (2021) claim that up to 48 bird and mammal extinctions have been prevented by conservation efforts since the UN

Convention on Biological Diversity came into force in 1993, and up to 2020 – which is a small number if we consider that 42,100 species are threatened with extinction, according to the International Union for the Conservation of Nature (2023).

The small number of successful reintroductions is due to different reasons, but Dolman *et al.* (2015) have argued that, unless animals in the wild are protected, captive breeding will not be significant. When a species is in danger of extinction in nature, it is not always a good idea to establish a breeding population in captivity, because without *in situ* conservation there is no sense in *ex situ* reproduction, that is, in captivity.

> Captive breeding can offer a last chance when species face imminent extinction, but ultimately depends on re-establishing a population in the wild. This has proved successful for some high-profile species, but in many cases it has not. Programmes can fail for many reasons, including delays in achieving successful breeding, failure to build up a self-sustaining population, domestication and loss of genetic diversity, and poor performance after releases into the wild. Captive breeding can reduce motivation and resources for conservation in the wild, with disastrous consequences. (Dolman, 2015, np)

In many cases, it is more complicated and expensive to maintain and reproduce animals in captivity than to protect their habitats and find ways to maintain a population that can reproduce in the wild. On the other hand, breeding animals in captivity under the pretext that they will be reintroduced can divert efforts and resources to protect them in the wild.

Animals raised in captivity present a series of problems that animals do not have in the wild. Among the main problems is the lack of genetic diversity. Even if animals from other zoos are used, the populations of animals in captivity are usually small; a reduced genetic pool forces genetic-related individuals to cross and reaches a point where inbreeding makes animal reproduction and survival difficult. This is one of the reasons why infant mortality in captive animals is much greater than that of the same animals in the wild. Boakes *et al.* (2007) investigated the effects of inbreeding in 119 zoo populations (covering 88 species of mammals, birds, reptiles and amphibians) and showed that inbreeding depression for neonatal survival was significant in all of them. Additionally, if individuals who come from these breeding programmes are being reintroduced for conservation purposes, this casts doubt on their chances of adaptation and survival.

Captivity and Animal Rights

There is another line of argumentation about the moral justification of zoos that focuses on the moral value, and rights, of animals. Animals do not have rights that derive from the political or social arrangements of humans, such as the right to vote or marriage, but there are rights that human beings have, and given the similarities between the physical and cognitive abilities of humans and animals, we cannot justifiably deny them to the latter if we recognize them for the former (Rachels, 1997; Regan, 2004). The rights to life and freedom are among the rights that human and non-human animals share. Someone has a right to life when she is aware of her own existence, values it and wants it to continue; this is particularly so when one has the ability to plan and anticipate her own future. Then we say that that individual has a fundamental interest in her own life that is protected by a right. On the same basis we can attribute to animals the right to life. If humans have the right to freedom, it is because without it our interests would be damaged in a significant way – freedom allows us to realize our capabilities and lead our lives according to our plans. For an individual to have an interest in freedom 'it is not necessary more than that the interference of other people causes harm by preventing them from doing what they want' (Giroux, 2016, p. 33). The captivity of animals is a violation of their right to freedom. By definition 'captivity' is the deprivation of freedom and it is a defining idea of zoos as we have traditionally known them. Zoos damage the vital interests of animals by keeping them in captivity. This claim should be qualified by recognizing that different species have different vital interests according to their needs: the lion's interest in a life of freedom that allows him to develop his abilities is not the same as that of a folivore, which needs less space to develop them. Both have the right to freedom, although their needs for freedom are different. However, many zoo animals of different species have tried to escape, as has been narrated by Jason Hribal in his book *Fear of the Animal Planet* (2011; see also Marris, 2021).

Animals in captivity have been deprived of the opportunity to behave naturally. They are subtracted from their natural habitat, separated from their herds and important links are broken for them. They are prevented from interacting with members of other species, sometimes even with members of their own species, or they are forced to live in abnormal social groups (such as only males or only females). Solitary species are forced to live in a group or in densities that generate social conflicts, they are prevented from procuring their own food and, in general, they are prevented from performing the behaviours that they would carry out in their natural habitats. Therefore, even in zoos with the best conditions, their right to freedom is violated.

Additionally, captivity is associated with various pathologies, both physical and psychological, developed by zoo animals. The mere exhibition of the animals to the public generates stress (Davey, 2007), not to mention the way they are captured, transferred and managed inside the zoo. Even though many zoos have stopped showing animals in steel-barred concrete cages, to make way for exhibitors trying to reproduce the natural habitat of the animals, many find themselves in places with climatic conditions for which the animals are not adapted. In addition, animals usually spend two-thirds of their lives locked in night houses, while the zoo is closed. Night houses are usually cages that lack any kind of environmental enrichment. Without environmental stimulation or interaction with other members of their species, the life of zoo animals tends to be boring and lead them to develop anxiety or depression and ultimately stereotypical behaviours, such as felines pacing and circling back and forth inside their exhibit, elephants' head-bobbing or polar bears that compulsively swim in circles (see Swaisgood and Shepherdson, 2005; the articles collected in Mason and Rushen, 2006). There are cases of neurosis caused by confinement. This leads some animals to show aggressive behaviours that they would not have in the wild. This shows that the vital interests of the animals are affected by captivity.

By claiming that animals' captivity can have negative effects on their welfare, I do not want to give the impression that freedom is valuable for animals because it allows them to obtain other things that are valuable to them, such as well-being. If this were so, then humans could continue to use and keep them captive as long as they were well and did not suffer. The interest of animals for freedom is not subordinated to their welfare; they have a direct interest in freedom to the extent that they are harmed when human interference in their lives prevents them from doing what they want to do. (Giroux (2016) argues against a positive conception of freedom (understood as a sophisticated version of autonomy) according to which humans could know better than animals what is in their best interest and keep them under their control. This happens when freedom is subordinated to welfare. Instead, she argues that both under a negative conception and under a republican conception of freedom, animals have a direct interest in being free and that their interest must be protected by recognizing their right to freedom.)

The situation of the right to life in zoos is as much or more serious. Breeding programmes tend to create a surplus of animals that zoos cannot maintain afterwards. For a zoo, it is enough to have only one male that can reproduce with the females of the herd. Having more than one can create rivalries and problems within the group, in addition to representing a maintenance expense. However, instead of stopping the reproduction of animals, zoos reproduce them because contraception can involve medical risks, because some animals can become infertile if they do not reproduce and also because, since they deprive them of a good part of their natural behaviours, it is important to let them reproduce and raise their offspring (Parker, 2017) – which is also a way to keep the group of animals of that species genetically robust. This creates surpluses (mostly males) and zoos have three solutions to get rid of them. The first is that they usually have exchange programmes with other zoos to avoid these problems but also possible endogamy. But on the one hand, moving from one zoo to another breaks social relations within the group, creates social instability and generates tensions for the transferred animal. Some animals end up going through several zoos during their lifetime. In addition, many zoos no longer need more males of species that are overrepresented. The second solution for surplus animals is to sell them to individuals, private institutions (which do not always have the necessary conditions for maintaining the animal) or hunting clubs (Humane Society of the United States, 2023). The third is killing (culling) or 'euthanizing' mostly healthy animals but useless for breeding programmes (they are 'genetically unnecessary') and expensive for the zoo. Lacy (1995), of the Chicago Zoological Park, has defended this option, arguing that this is preferable to sending them to lower quality zoos, where animals will have a bad time. The European Association of Zoos and Aquariums does not publish information on the number of healthy animals that are killed, but its executive director, Lesley Dickie, estimated that between 3000 and 5000 animals are killed in European zoos per year (Barnes, 2014). The Copenhagen Zoo (where the giraffe Marius was 'euthanized' in 2014) kills between 20 and 30 animals per year, usually antelopes, reindeer, goats, but also lions, tigers, zebras and bears (Parker, 2017). Although zoos in the USA prefer to transfer animals, AZA regulations also allow killing and it is a common practice, especially among less charismatic animals (Parker, 2017). The reintroduction into the wild is usually not considered in cases of surplus animals because they often do not adapt, but mainly because it is very expensive for the zoo and because in many cases their natural habitats have disappeared. In any case, the killing of healthy animals in zoos is not ethically justifiable if their right to life is recognized.

The recognition of animals' rights is incompatible with the existence of zoos. Someone may object that rights may be limited when there are other rights in conflict or when there are strong reasons of social utility. But neither the entertainment, education, research nor conservation carried out in zoos justify the limitation of the fundamental rights of animals. Especially when these justifications tend to be nothing more than the expression of good intentions without a solid foundation to support them.

Moving Forward

It is clear that zoos are not going to disappear overnight, but there are ways in which they can be transformed so that the animals that live there are in better conditions. First, there should be more rigorous legislation preventing the destruction of the natural habitats of many species of animals. Second, zoos should look for ways to send their surplus animals to sanctuaries or places where they can be integrated into real conservation programmes. Zoos must also stop buying animals – as well as selling them to laboratories, circuses, private collectors or hunting ranches, as happens in some zoos (Clifton, 1988). Above all, they must stop extracting animals from the wild, such as reptiles, amphibians and marine fauna (for example, orcas and dolphins that are used in aquariums). Finally, we must find a way for zoos to become ecological parks or sanctuaries. A sanctuary is a facility where animals are taken to live and be protected throughout their lives; unlike animal shelters, sanctuaries do not seek to place them with individuals or institutions, but keep them until their natural death. An example of a zoo that is fulfilling the role of sanctuary is the monkey rescue centre Monkey World, in the UK, where many primates that had been abused in circuses or laboratories are kept. The Elephant Sanctuary in Tennessee has also received elephants from circuses and zoos; this organization has several hundred acres of forests and pastures where elephants can live the rest of their lives in freedom. There are many other similar sanctuaries registered with the Global Federation of Animal Sanctuaries, a non-profit organization, founded in 2007, that certifies sanctuaries, and rescue and animal rehabilitation centres around the world. In 2016, the Buenos Aires Zoo closed its doors to become an eco-park, sending many of its animals to sanctuaries and keeping only animals that, due to age or disease, could no longer be transferred.

Animal sanctuaries are becoming a viable alternative for the transformation of zoos and for the conservation of wildlife, especially in those cases where their natural habitats have been lost. The idea will become stronger when people who visit zoos begin to recognize the rights of animals, begin to wonder whether it is right to keep thousands of animals captive, question the reasons commonly given by zoos, stop visiting them and start supporting sanctuaries. These should be places where animals have the opportunity to exercise their rights to freedom and life, as well as other rights. Only then we will begin to repair the historical debt we have to animals.

References

Andersen, L.L. (2003) Zoo education: from formal school programmes to exhibit design and interpretation. *International Zoo Yearbook* 38(1), 3875–3881. DOI: 10.1111/j.1748-1090.2003.tb02066.x.

Anderson, U.S., Kelling, A.S. and Maple, T.L. (2008) Twenty-five years of zoo biology: a publication analysis. *Zoo Biology* 27(6), 444–457. DOI: 10.1002/zoo.20177.

Association of Zoos and Aquariums (2019) AZA zoos and aquariums contribute $24 billion to U.S. economy. AZA News Releases. Available at: https://www.aza.org/aza-news-releases/posts/aza-zoos-and-aquariums-contribute-24-billion-to-us-economy-?locale=en (accessed 27 June 2022).

Association of Zoos and Aquariums (2020) Annual Report on Conservation and Science 2020. Available at: https://www.aza.org/annual-report-on-conservation-and-science (accessed 25 April 2023).

Association of Zoos and Aquariums (2023a) Research and Science. Available at: https://www.aza.org/research-and-science (accessed 30 January 2023).

Association of Zoos and Aquariums (2023b) Conservation. Available at: https://www.aza.org/conservation (accessed 30 January 2023).

Association of Zoos and Aquariums (2023c) AZA Accredited and Certified Members. Available at: https://www.aza.org/inst-status?locale=en (accessed 30 January 2023).

Association of Zoos and Aquariums (2023d) Species Survival Plan® Programs. Available at: https://www.aza.org/species-survival-plan-programs?locale=en (accessed 30 January 2023).

Baker, A. (2007) Animal ambassadors: an analysis of the effectiveness and conservation impact of *ex situ* breeding efforts. In: Zimmermann, A., Hatchwell, M., Dickie, L. and West, C. (eds) *Zoos in the 21st Century: Catalysts for Conservation?* Cambridge University Press, Cambridge, UK, pp. 139–154.

Barnes, H. (2014) How many healthy animals do zoos put down? BBC News, 27 February. Available at: http://www.bbc.com/news/magazine-26356099 (accessed 2 February 2023).

Beck, B.B., Rapaport, L.G., Price, M.R.S. and Wilson, A.C. (1994) Reintroduction of captive-born animals. In: Olney, P.J.S., Mace, G.M. and Feistner, A.T.C. (eds) *Creative Conservation: Interactive Management of Wild and Captive Animals*. Chapman and Hall, London, pp. 265–286. DOI: 10.1007/978-94-011-0721-1.

Berger, J. (1992) Why look at animals? In: *About Looking*. Vintage, London, pp. 1–26.

Boakes, E.H., Wang, J. and Amos, W. (2007) An investigation of inbreeding depression and purging in captive pedigreed populations. *Heredity* 98(3), 172–182. DOI: 10.1038/sj.hdy.6800923.

Bolam, F.C., Mair, L., Angelico, M., Brooks, T.M., Burgman, M. *et al.* (2021) How many bird and mammal extinctions has recent conservation action prevented? *Conservation Letters* 14(1), e12762. DOI: 10.1111/conl.12762.

Cain, L. and Meritt, D. (2007) The demand for zoos and aquariums. *Tourism Review International* 11(3), 295–306. DOI: 10.3727/154427207783948856.

Carr, N. and Cohen, S. (2011) The public face of zoos: images of entertainment, education and conservation. *Anthrozoös* 24(2), 175–189. DOI: 10.2752/175303711X12998632257620.

Clarke, A.G. (2009) The Frozen Ark project: the role of zoos and aquariums in preserving the genetic material of threatened animals. *International Zoo Yearbook* 43(1), 222–230. DOI: 10.1111/j.1748-1090.2008.00074.x.

Clayton, S., Fraser, J. and Saunders, C.D. (2008) Zoo experiences: conversations, connections, and concern for animals. *Zoo Biology* 28(5), 377–397. DOI: 10.1002/zoo.20186.

Clifton, M. (1988) Chucking zoo animals overboard: how and why Noah culls the ark. *The Animals' Agenda* 8, 14–22.

Davey, G. (2007) Visitors' effects on the welfare of animals in the zoo: a review. *Journal of Applied Animal Welfare Science* 10(2), 169–183. DOI: 10.1080/10888700701313595.

Dolman, P.M. (2015) Critically endangered species should be left to breed in the wild. *Phys.org*. Available at: https://phys.org/news/2015-06-critically-endangered-species-left-wild.html (accessed 1 February 2023).

Dolman, P.M., Collar, N., Scotland, K.M., Burnside, R.J. and Bode, M. (2015) Ark or park: the need to predict relative effectiveness of *ex situ* and *in situ* conservation before attempting captive breeding. *Journal of Applied Ecology* 52(4), 841–850. DOI: 10.1111/1365-2664.12449.

Dunlap, J. and Kellert, S.R. (2004) Animal welfare and rights: zoos and zoological parks. In: Post, S.G. (ed.) *Encyclopedia of Bioethics*, 3rd edn, Vol. 1. Macmillan, New York, pp. 208–212.

Escribano, N., Ariño, A., Pino-del-Carpio, A., Galicia, D. and Miranda, R. (2021) Global trends in research output by zoos and aquariums. *Conservation Biology* 35(6), 1894–1902. DOI: 10.1111/cobi.13757.

Evans, W. (2015) Posthuman rights: dimensions of transhuman worlds. *Teknokultura* 12(2), 373–384. DOI: 10.5209/rev_TK.2015.v12.n2.49072.

Frost, W. (2011) Rethinking zoos and tourism. In: Frost, W. (ed.) *Zoos and Tourism. Conservation, Education, Entertainment?* Channel View, Bristol, UK, pp. 1–8. DOI: 10.21832/9781845411657.

Giroux, V. (2016) Animals do have an interest in liberty. *Journal of Animal Ethics* 6(1), 20–43. DOI: 10.5406/janimalethics.6.1.0020.

Godinez, A.M. and Fernandez, E.J. (2019) What is the zoo experience? How zoos impact a visitor's behaviors, perceptions, and conservation efforts. *Frontiers in Psychology* 10, 1746. DOI: 10.3389/fpsyg.2019.01746.

Gusset, M. and Dick, G. (2011) The global reach of zoos and aquariums in visitor numbers and conservation expenditures. *Zoo Biology* 30(5), 566–569. DOI: 10.1002/zoo.20369.

Hribal, J. (2011) *Fear of the Animal Planet: The Hidden History of Animal Resistance*. AK Press, Oakland, California.

Humane Society of the United States (2023) Captive hunts fact sheet. Available at: https://www.humanesociety.org/resources/captive-hunts-fact-sheet (accessed 1 February 2023).

International Journal of Heritage Studies (2008) Zoos as heritage. Special issue. *International Journal of Heritage Studies* 14, 1. DOI: 10.1080/13527250701712398.

International Union for the Conservation of Nature (2023) IUCN Red List. Available at: http://www.iucnredlist.org (accessed 1 February 2023).

Jamieson, D. (2004a) Against zoos. In: *Morality's Progress*. Oxford University Press, New York, pp. 166–175.

Jamieson, D. (2004b) Zoos revisited. In: *Morality's Progress*. Oxford University Press, New York, pp. 176–189.

Jensen, E. (2014) Evaluating children's conservation biology learning at the zoo. *Conservation Biology* 28(4), 1004–1011. DOI: 10.1111/cobi.12263.

Kisling, V. (ed.) (2001) *Zoo and Aquarium History: Ancient Animal Collections to Zoological Gardens*. CRC, Boca Raton, Florida. DOI: 10.1201/9781420039245.

Kreger, M. (1998) Zoo visitor-animal interactions. In: Bekoff, M. and Meaney, C.A. (eds) *Encyclopedia of Animal Rights and Animal Welfare*. Greenwood Press, Westport, Connecticut, pp. 378–380.

Lacy, R. (1995) Culling surplus animals for population management. In: Norton, B.G., Hutchins, M., Stevens, E.F. and Maple, T.L. (eds) *Ethics on the Ark. Zoos, Animal Welfare, and Wild Conservation*. Smithsonian Institution Press, Washington, pp. 187–194.

Linke, S. and Winter, C. (2011) Conservation, education or entertainment: what really matters to zoo visitors? In: Frost, W. (ed.) *Zoos and Tourism. Conservation, Education, Entertainment?* Channel View, Bristol, UK, pp. 69–81. DOI: 10.21832/9781845411657.

Loh, T.-L., Larson, E.R., David, S.R., de Souza, L.S., Gericke, R. *et al.* (2018) Quantifying the contribution of zoos and aquariums to peer-reviewed scientific research. *Facets* 3(1), 287–299. DOI: 10.1139/facets-2017-0083.

Ludwig, E.G. (1981) People at zoos: a sociological approach. *International Journal for the Study of Animal Problems* 2, 310–316.

Marcellini, D.L. and Jenssen, T.A. (1988) Visitor behavior in the national zoo's reptile house. *Zoo Biology* 7(4), 329–338. DOI: 10.1002/zoo.1430070404.

Margodt, K. (2010) Zoos as welfare arks? Reflections on an ethical course for zoos. In: Acampora, R.R. (ed.) *Metamorphoses of the Zoo. Animal Encounter after Noah*. Lexington Books, Lanham, Maryland, pp. 11–36.

Marris, E. (2021) Modern zoos are not worth the moral cost. *The New York Times,* June 11.

Mason, P. (1999) Zoos as heritage tourism attractions: a neglected area of research? *International Journal of Heritage Studies* 5(3–4), 193–202. DOI: 10.1080/13527259908722267.

Mason, G. and Rushen, J. (eds) (2006) *Stereotypic Animal Behaviour: Fundamentals and Applications to Welfare*. CABI, Wallingford, UK. DOI: 10.1079/9780851990040.0000.

Moss, A. and Esson, M. (2013) The educational claims of zoos: where do we go from here? *Zoo Biology* 32(1), 13–18. DOI: 10.1002/zoo.21025.

Parker, I. (2017) The Culling. *The New Yorker*, January 16, 43–53.

Patrick, P.G. and Tunnicliffe, S.D. (2013) *Zoo Talk*. Springer, Dordrecht, the Netherlands. DOI: 10.1007/978-94-007-4863-7.

Patrick, P.G., Matthews, C.E., Ayers, D.F. and Tunnicliffe, S.D. (2007) Conservation and education: prominent themes in zoo mission statements. *The Journal of Environmental Education* 38(3), 53–60. DOI: 10.3200/JOEE.38.3.53-60.

Rachels, J. (1997) Do animals have rights? In: *Can Ethics Provide Answers*. Rowman and Littlefield, Lanham, Massachusetts, pp. 81–98.

Regan, T. (2004) *The Case for Animal Rights*, 2nd edn. University of California Press, Berkeley and Los Angeles, California.

Sommer, R. (1972) What do we learn at the zoo? *Natural History* 81, 26–29 and 84–85.

Swaisgood, R.R. and Shepherdson, D.J. (2005) Scientific approaches to enrichment and stereotypies in zoo animals: what's been done and where should we go next? *Zoo Biology* 24(6), 499–518. DOI: 10.1002/zoo.20066.

Tribe, A. (2004) Zoo tourism. In: Higgenbottom, K. (ed.) *Wildlife Tourism: Impacts, Management and Planning*. Common Ground, Melbourne, Australia, pp. 35–36.

Wharton, D. (2007) Research by zoos. In: Zimmermann, A., Hatchwell, M., Dickie, L. and West, C. (eds) *Zoos in the 21st Century: Catalysts for Conservation?* Cambridge University Press, Cambridge, UK, pp. 178–191.

Zimmermann, A. and Wilkinson, R. (2007) The conservation mission in the wild: zoos as conservation NGOs? In: Zimmermann, A., Hatchwell, M., Dickie, L. and West, C. (eds) *Zoos in the 21st Century: Catalysts for Conservation?* Cambridge University Press, Cambridge, UK, pp. 303–321.

4 Fill Your Cup: A Multi-species Exploration of Balinese Kopi Luwak Agrotourism

Jes Hooper*
University of Exeter, Exeter, UK

Abstract

Kopi luwak (Indonesian for 'civet coffee') is known as the world's most expensive coffee. It is made from coffee beans that have been partially digested by the Asian palm civet (*Paradoxurus hermaphroditus*), a small cat-like mammal native to Southeast Asia. Since achieving international fame in the early 2000s, kopi luwak tourism has emerged in curated sites across Bali where tourists can see the process of kopi luwak production. This chapter explores the natureculture dynamics of the multi-species interactions that occur in Balinese kopi luwak agrotourism by applying a posthumanism lens to the content analysis of 3364 TripAdvisor reviews posted between 2010 and 2020. Through posthumanism, this chapter demonstrates the ways in which, through their exploitation and labour, both indigenous humans and animals transgress from beings with intrinsic worth and individual agency to display objects in service of the tourist's gaze and economic interests. Overall, this chapter reveals the inauthenticity of kopi luwak agrotourism, its success reliant instead on the tourist's idea of cultural heritage and pristine nature.

Introduction

Kopi luwak (Indonesian for 'civet coffee') is known as the world's most expensive coffee. It is made from coffee beans that have been partially digested by the Asian palm civet (*Paradoxurus hermaphroditus*), a small cat-like mammal native to Southeast Asia. According to legend, kopi luwak is the rarest coffee on Earth because farmers rely on wild civets entering coffee plantations at night where they feed on the ripest cherries before depositing the newly transformed beans in their scat. In essence, kopi luwak is an indigenous product, one that is created through the digestive tract of local civets and the hands of local people who pick the prized beans from the animal's excrement before cleaning, drying and roasting them (Marcone, 2004). But kopi luwak is also steeped in colonial histories. Over 300 years ago, Indonesian coffee farmers under Dutch colonial rule were forbidden to use coffee from their own crop and so instead turned to the waste product of marginalized animals (Muzaifa *et al.*, 2019). Kopi luwak only recently became fetishized within economically advanced nations, featuring on the Oprah Winfrey show in 2003 and in the 2007 film *The Bucket List* starring Jack Nicholson and Morgan Freeman.

However, with kopi luwak's rising demand has come a shift in the role of its creator. No longer permitted visitors to coffee farms, civets

*jh1220@exeter.ac.uk

 Tourism, Heritage and Commodification of Non-human Animals (eds A. López López *et al.*)
DOI: 10.1079/9781800623309.0004

are now seen as natural capital and natural heritage and increasingly captured and caged for mass kopi luwak production. Despite relying on the false claims of a global availability of only 127 kg per year (Marcone, 2004), kopi luwak is now found in abundance in the international coffee market, where it continues to reach prices of up to US$260 per kilo (Muzaifa *et al.*, 2018). While the exact volume is unknown, the global kopi luwak market's net worth is set to reach US$10 billion by 2030 (Straits Research, 2022). Its fame has also seen the emergence of kopi luwak tourism.

Nestled amid the trees overlooking the juxtaposed Balinese rice paddies and the golf courses of the popular 'tourist trail', giant statues are being erected of the Asian palm civet, effigies to ground the history of kopi luwak discovery into the contemporary Balinese landscape. Here, tourists are invited to witness the re-creation of traditional kopi luwak production, and meet coffee plants, native civets and village elders before partaking in the taste of kopi luwak in its unique place of origin. Yet it is here that authenticity and heritage are in fact falsified through modes of multi-species exploitation.

As an emergent tourist industry, kopi luwak agrotourism has been poorly studied, regardless of its growing popularity since 2010. While the few studies conducted so far illustrate the animal welfare issues associated with kopi luwak production and tourism (Carder *et al.*, 2016; Lachenmeier and Schwarz, 2021; Hooper, 2022a), it remains unclear how tourists perceive kopi luwak as a gastronomic experience, one that relies on the interplay between humans, plants and animals. By analysing the perceptions of tourists who have visited kopi luwak agrotourism sites, and exploring the ways in which they perceive and engage with the attractions on offer, this chapter aims to illustrate the ways animals are exploited in the curation of heritage. It is my hope that such insights will help inform and inspire a posthumanism tourist ethic by contending the human–animal divide (Cohen, 2019). This chapter utilizes TripAdvisor to develop a multi-species virtual ethnography of Balinese kopi luwak agrotourism, drawing on content analysis of 3364 tourist reviews written by visitors to 25 kopi luwak tourist sites between 2010 and 2020.

Methods

First, TripAdvisor was used as a search engine (using the terms 'kopi luwak', 'civet', 'coffee' and 'civet coffee') to identify sites with live civets (denoted by a paw print in the description) and to collect tourist reviews for each location. Tourist reviews were analysed using grounded theory in which concepts were generated from the data. First, each review was read for initial data familiarity and identification of emergent themes (which were grouped into thematic categories), before the reviews were read again and scored against each category. Further details on the thematic categorization process are available in Hooper (2022a). Second, in addition to the descriptions, videos and photographs included on TripAdvisor, participant interviews were conducted with three holidaymakers who had attended these sites. These participants had responded to a call for interviewees published online (https://www.thecivetproject.com; accessed July 2023). While there was no incentive for participation, each participant had found the call on Google when researching civet coffee after their visit to Bali and wished to share their experience. Interviews were conducted via Zoom and participants were asked to narrate their tour experience.

Animal-derived Gastronomic Heritage: Multi-species and Posthuman Perspectives

Just as 'tastescapes' are reserved for foods secured within the cultural heritage of a particular country or region, such as pasties in Cornwall or Port wine in Porto (Bruncevic and Linne, 2018), so too is kopi luwak marketed as a tastescape of the Balinese tourist experience. Tastescapes are said to provide a sensory experience that goes beyond the visual bias of traditional tourism (Urry, 2002), though the emergence of kopi luwak tourism has certainly benefited from integration into an already well-established global tourist hotspot. The beauty of the Balinese landscape has been a tourist draw for decades and in the 1970s designated tourist routes were installed through the rice fields to enable access to the most rewarding views

(Cole, 2012). When planned effectively, rural tourism has been known to afford communities the opportunity to safeguard environmental biodiversity, local economy and tradition (Nicolaides, 2020). Agrotourism has thus become a popular tourist strategy throughout Indonesia, promoting additional income to farming communities via product and service diversification and sustainable income (Hakim *et al.*, 2019).

Marketed as kopi luwak agrotourism, such sites are small in scale and located in difficult terrain on the island's main tourist trail, extending from Denpasar International Airport in the southern peninsula to Mount Batur to the northwest of Mount Agung, making them unsuitable locations for genuine kopi luwak production. In fact, the majority of kopi luwak production for local and global markets takes place in areas far from the tourist trail, locations that Cahill (2020) likens to Tsing's (1993) marginalized 'out-of-the-way places'. Kopi luwak tourism venues, therefore, would be best described as sites of curated authenticity rather than sites of genuine agrotourism as local humans, plants and animals are entwined in an assimilated representation of traditional narrative, a contact zone (Haraway, 2008) fit for the expectations of holidaymakers.

While the concept of 'authenticity' has been widely debated within tourism studies (e.g. Belhassen and Caton, 2006; Reisinger and Steiner, 2006), the theory of staged authenticity predates postmodernism (Baudrillard, 1981) and can be observed in many tourist activities from re-enactments (Lovell and Hitchmough, 2020) to film tourism (Buchmann *et al.*, 2010). Similar findings are observed in kopi luwak tourism, whereby each plantation is constructed in the same way, so that tourists are encouraged to believe they are witnessing kopi luwak production in a legitimate site of creation. First, tourists are greeted at the entrance by a tour guide who guides them through a small garden of, often labelled, indigenous plants and cash crops such as coffee. During this stage, tourists are invited to view the coffee cherries in their natural form, and they are shown the difference between male and female plants. They are then guided towards a row of metal cages, commonly fitted with slatted or wire floors and a bowl of coffee cherries, each occupied by one or two civets (see Fig. 4.1). Here the tour group can view how the unique coffee is 'processed' through civet consumption and defecation. Following on in the production line, the tour then stops at a grinding station to observe an elderly Balinese woman roasting coffee beans on an open fire. Tourists are encouraged to try grinding the roasted beans using a traditional stone pestle and mortar, before being offered complimentary endemic teas and coffees with the option to purchase kopi luwak for a nominal fee (typically US$5). The entire curated experience usually lasts no longer than 30 min before the tourists exit through a gift shop.

NatureCulture Framework

As with many tropical tourist destinations, Bali is marketed as a place rich in natural beauty and culture. Tourists are enticed with images of sandy beaches framed by palms, dappled light entering jungle canopies and ancient temples nestled under volcanic backdrops. According to environmentalist E. O. Wilson, humans have an intrinsic tendency to seek interactions with nature, a term he dubs as the 'biophilia hypothesis' (Kellert and Wilson, 1993). With 55% of the world's human population now living in urban environments (UN, 2018) and a growing ease of access to international travel (Glaesser *et al.*, 2017), it may be unsurprising that holidaymakers are increasingly drawn to exotic naturalistic environments. Yet the cosmopolitan globalized ecologies experienced today challenge the dualism promoted by the biophilia hypothesis whereby humans are observed as separate to nature (Fuentes, 2010). Instead, the natureculture rubric provides an appealing alternative, addressing the synonymy between nature and culture while challenging the dualism instilled by a discourse of human exceptionalism (Haraway, 2003).

The reflective ethnographic approach offered by a natureculture framework allows for a more inclusive and nuanced assessment of multi-species encounters to better understand the ways in which inter-species ecologies shape societal practices. By viewing human–animal interactions in a natureculture framework, ethnography gains further depth by recognizing the entwined relations between behavioural and

Fig. 4.1. An Asian palm civet (*Paradoxurus hermaphroditus*) sits in an empty cage in a Balinese kopi luwak tour. (Anonymous TripAdvisor reviewer.)

ecological processes and religious, economic and political factors. Fuentes' (2010) ethnoprimatology essay on the multi-species interface between Balinese people and macaques within rural and tourist settings eloquently demonstrates the benefit of such an approach. Through a natureculture lens, the livelihoods of both humans and alloprimates were shown to be inextricably linked through mutual ecologies of politics, environment, economy, biology and culture. Both human and animal were considered as active participants in a multi-species encounter. By applying a natureculture lens to kopi luwak tourism, this chapter offers a posthumanism critique of human–civet interactions in this emergent tourist industry.

A natureculture framework is highly applicable to kopi luwak tourism, an industry built on multi-species entanglements where civets, coffee plants, the natural landscape and humans are deeply connected. As with all agrotourism enterprises, the sustainability of kopi luwak tourism relies on natural resources and biodiversity, on which both the product (kopi luwak) and the attraction (plantation landscapes) rely. The following sections explore common natureculture themes presented by tourist reviews written on the TripAdvisor website.

A Desire to Consume the Traditional Balinese Landscape

While the designated tourist routes through the rice fields date back to the 1970s, the subsequent development of tourist facilities has injected an anthropogenic footprint of economically advanced nations on the traditional Balinese landscape. Tourist facilities have had a detrimental effect on the local ecosystem as hotels, golf courses and rice fields compete for water to sustain them (Cole, 2012). The juxtaposition between tourist desires and environmental needs has led some scholars to comment on the lack of awareness held by tourists regarding the cultural and ecological stress felt by Balinese landscapes (Collins-Kreiner and Wall, 2015). Indeed, kopi luwak reviewer comments reflected this mentality. One tourist noted:

> The ambience was great except there was a new building construction seen on the rice terrace that disturbed the natural view… If only they had the rice terrace all natural.

Reviews such as this showed that tourists had a lack of understanding regarding the role of tourism in the construction of buildings in rural Balinese settings. The associated displeasure came from the tourist's own desire to consume the traditional Balinese landscape rather than concern for the environment or local inhabitants (humans, plants or animals). Furthermore, describing rice paddies as 'natural' showed a lack of awareness of the cultivated nature of monocrop plantations (Cole, 2012). The rise in 'selfie-culture' was a big draw to visiting tourists and buildings were considered by many to detract from the pristine exotic nature they sought to purvey on social media.

> Great place for a selfie. Too bad though, someone is building some major concrete accommodation across the valley so don't really expect this nice view to last long.

Unsurprisingly then, for most positively ranked reviews, it was the ability for tourists to share their own personal story of immersion in the iconic Balinese landscape (pristine jungle combined with traditional rice paddies) which was of primary interest rather than civet coffee.

The importance of virtual storytelling for the tourist experience was capitalized on through the development of photo souvenir opportunities by kopi luwak agrotourism enterprises. The natural aesthetics of the surrounding plantation environment were curated in such a way as to promote the fantasized reality of full immersion in tropical Balinese nature. Jungle swings, hanging platforms and verandas with panoramic views were present in most sites, whereby tourists could pay an additional fee to have photographs taken amid the iconic Balinese backdrop, photos which inevitably went on to be shared on social media. The sharing of such imagery on social media sites facilitates the visitors' sense of 'neoliberal branded-self' (Smith, 2018, p. 10), a form of social capital whereby both the visitor's body and geographic location become sites of commodification. Where the images act as social capital in the form of 'likes' and 'shares', these images also offer potential financial capital for the kopi luwak tourist operators by way of serving as electronic-word-of-mouth advertisement.

A Desire to Reconnect with Nature through Coffee Production and Consumption

Coffee itself is inarguably entwined in nature-culture processes, as the chain of production connects local and global landscapes, economic systems and natural resources (Tucker, 2017). The chain of connection from bean to cup offers a cosmopolitan experience to the consumer as the central link connecting these global chains. The ubiquitous nature of coffee in societies across the globe combined with heavily processed goods can however proliferate a false dualism between human and nature, as reviewers note their separation from coffee as a natural product, seeing it instead as a human-made commodity. One reviewer observed:

> The house of coffee is a small but informative tour through the coffee making process, most of which I didn't know of, despite guzzling coffee daily for years.

Many tourists claimed to find the display of indigenous plants to be educational, most reviews included generalized statements of interest, such as 'It's interesting to see all the spice, fruit and coffee trees', whereas some reviewers provided self-reflection indicative of a sense of separation from nature. One reviewer claimed it was 'great fun to see all the spices I buy in dried format in North America as real plants growing in Indonesia'. The supposition that store-bought products were no longer 'real plants' highlights the perceived dichotomies between humans/nature and nature/culture instilled by globalized trade in natural resources. Agrotourism therefore granted tourists an opportunity to reconnect with nature through the raw form of the products they engage with on a daily basis.

The Perception of Supporting Authentic Agricultural Heritage

With tourism accounting for 10.5% of Indonesia's total employment (OECD Tourism Statistics, 2022), fewer Balinese people now rely on agriculture for their livelihoods. Thus, as agriculture makes way for tourist culture in Bali, the perceived authentic agricultural practices tourists wish to support are effectively becoming rarer through tourist activity. Yet, the perceived authenticity within the natureculture experience is cemented by the term 'agrotourism' and 'plantation' in many of the site names (such as 'Ubud Sari Coffee Luwak Agrotourism' and 'Luwak Coffee Bali Plantation'), terms that lead tourists to believe they are visiting working plantations, supporting rural communities and helping to preserve ecosystems. One reviewer explained:

> Agrotourism is now the 'IN' thing of presenting and selling local products. It's not to mislead tourists but rather to give them a glimpse of the how the original plants look like, how it's grown, how tediously the products are made i.e. Roasting the coffee for several hours before grinding them the old way, that is, pounding them instead of how Starbucks ground them with machines. It's cool ... especially for kids. One can probably equate it with the serene tea ceremony of the Japanese demonstrating the exquisite simplicity of the entire process.

Curated demonstrations of traditional coffee processing where the coffee is roasted on a fire and ground by hand, almost exclusively by an elderly Balinese woman, was considered by many reviewers to be a traditional practice and highly valued as a cultural and educational experience. Reviewers tended to romanticize their perceived cultural encounter with comments, such as: 'A Balinese native woman roasting coffee by hand – she kindly let me sit with her and stir the coffee beans' and 'I was mesmerized by this sweet old woman roasting the coffee beans. She has the sweetest face you will ever see'.

Reviewer descriptions of the coffee production process were often reflective of colonial discourses. In applying language such as 'native' and 'sweet' users reinforced a historically engrained power dynamic, framing the local people as what Smith depicts as 'genericized icons of exoticism that serve to imbue the tourist's experience with authenticity' (2018, p. 4). The claim to be 'mesmerized by' plantation workers further emphasized the object nature of those working within the exhibit. In such cases, tourists appeared to believe the traditional kopi luwak production narrative and therefore did not recognize the labour of these women, both in terms of their employment as display items within the tour site and their coffee roasting efforts. One review highlights the juxtaposition between human and animal as displayed objects: 'You also meet a lovely old lady roasting coffee beans, as well as the civets used in the coffee bean fermentation process along the way'.

Such reviews highlight the ways in which animals and women transform from living beings with agency, to objects in service to the tourist gaze.

Kopi Luwak as a Perceived Tastescape of Bali

The presentation of kopi luwak as a tastescape of Bali was a common theme within tourist reviews, as civet coffee tasting was frequently depicted as a 'traditional' and 'local' experience, one which was romanticized in many cases despite the product's colonial history. Indeed, as seen above, one reviewer likened kopi luwak tourism to Japanese tea ceremonies. Everett argues that food tourism has 'fostered new patterns of tourism consumption and the development of new individualized identities' (2008, p. 340). Not only is kopi luwak, and coffee more generally, intimately tied up with the cultural identity and heritage of Indonesia, the love for drinking coffee also featured as a form of personal identity among kopi luwak tourists. Reviewers frequently self-identified as 'coffee lovers' and 'coffee enthusiasts'. The chance to partake in a coveted coffee blend was often cited as a motivational factor towards attending their chosen plantation, as was the unique opportunity presented by its contextual affordability.

> It was $5 a cup (I've paid way more than that for coffee elsewhere) compared to $160 for a pound in other parts of the world.

The tourist experience therefore serves as a unique gateway into an elevated social class that many reviewers claimed was not within reach in their own countries. One reviewer claimed '55,000IDR sounds expensive but it's only $5.50'. The recognition of kopi luwak by international tourists as 'only $5.50' once again carries colonial narratives whereby the Indonesian identity is commodified as a cheap resource in the global marketplace (for similar findings in India, see Smith, 2018).

Often tourists perceive kopi luwak consumption in Bali as a once in a lifetime opportunity. The product's claimed heritage value was promoted throughout guided tours, reinforcing the idea that to refuse to partake would be a missed opportunity. The conceptual belief that kopi luwak is authentic to Bali confirms Bruncevic and Linne's (2018) claims that the 'tastescape' concept reinforces the fetishization of authenticity as something sublime, that can only occur in the place of origin. Furthermore, where the plantations offer an ocular-centric representation of curated natureculture in the form of jungles, mountains, rice paddies, wildlife and traditional coffee roasting practices, kopi luwak provides a taste experience that truly embodies a cross-species connect. Perhaps most striking is the embodiment of natureculture through the act of faecal commodification.

There are fewer connections closer to the natural world than the bodily process of ingesting, assimilating and expelling nature. Yet in many cultural contexts (to which many reviewers themselves belonged) bodily waste is considered disgusting, dangerous and to be avoided (Douglas, 2003). With plantations actively encouraging tourists to view civet scat (as demonstrated in Fig. 4.2), kopi luwak therefore dispels the societal motif surrounding natural bodily processes in challenge of dualistic ideas of nature separate to culture. When kopi luwak is sampled in its space of cultural heritage, tourists shed their own cultural identity to experience the destination as if they were one of the indigenous locals. Such immersive experiences are denoted by Smith as 'fantasized assimilation' (2018, p. 4). It is for this reason that kopi luwak tourism occupies a unique niche as a tastescape within Bali's tourist culture, one which now relies on the cultivation of native landscapes and the harvesting of native civets.

The Adoption of the Civet as an Icon of Heritage

Despite the story of kopi luwak's discovery dating back 300 years, the affiliation between kopi luwak and the island of Bali is an emergent phenomenon. The earliest kopi luwak agrotourism sites featured on TripAdvisor coincide with the rise of the product's fame in the West during the early 2000s. Its debut on the international stage, via a popular television talk show and Hollywood film in 2003 and 2007, was the likely catalyst to kopi luwak emergence as a tourist activity on the island in 2010. Indeed, its association with celebrity culture was a marketing ploy reiterated, and relied on, by the kopi luwak tourism sector. Unlike previous findings, which found positive correlations between celebrity endorsement and human–animal interactions (for the pet trade, see Nekaris *et al.*, 2013), few kopi luwak agrotourism reviewers acknowledged specific pop-culture references in their motivations for attending a site of kopi luwak agrotourism. One reviewer implied they had been previously unaware of the coffee's fame in Hollywood until their agrotourism visit, stating '*apparently* it features in [the movie] The Bucket List' (emphasis added). Most reviewers also claimed they had been directed to plantations by their driver or had attended as part of an organized tour, and so had not intended to visit the plantation based on its link to film or media. Kopi luwak's authenticity and fame is therefore mostly reinforced by the plantation guides, who directly emphasize civet coffee as 'the world's most expensive and most famous coffee' to their visitors. Similarly, the sites are constructed to promote narratives surrounding fame and celebrity status. Tourists pose in front of a Hollywood-style sign that sits in the foothills of one 'plantation' which reads: 'Bali Cat Poo Chino' (see Fig. 4.3).

Fig. 4.2. Civet scat pitted with partially digested coffee beans on display in a Perspex viewing box as part of a Balinese kopi luwak tour. (Anonymous TripAdvisor reviewer.)

Despite the industry's infancy as a tourism enterprise, the image of the civet is projected by agrotourism sites as a figure of monumental importance to the heritage of the surrounding landscape. Kopi luwak tourist sites have adorned their facilities with iconized statues of the wild animals. Civet statues, some of which stand over two stories tall (see Fig. 4.4), depict a powerful and dominating animal presence for tourists to look up to. Often painted with red eyes, sharp claws and a dominating presence, these impressive yet curious figures are a bold celebration of the power behind the lucrative Balinese kopi luwak business. Yet civet statues and the branding of kopi luwak as 'Cat-Poo-Chino' highlights an insidious aspect to the product's marketing strategy whereby the civet is represented in a way which detaches the lived experience of the animal from the product.

While the coffee is named after the animal from which it derives, civets are in fact an absent referent in the kopi luwak experience as their agency is negated along with a deferred moral responsibility for their well-being. Speaking in relation to the production of meat, Adams (2018) explains that animals are made absent both literally through their slaughter (the dead body replaces that of the living animal for meat to exist) and figuratively through language (the naming of animal bodies as 'meat' protects consumers from conjuring the images of animal corpses). In the case of civets, despite their

Fig. 4.3. Two tourists pose in front of a Hollywood-style sign 'Bali Cat Poo Chino' on the hillside overlooking a kopi luwak agrotourism site. (Anonymous TripAdvisor reviewer.)

naming in the product itself, their representation excludes the reality of their lived experience in the captive production of kopi luwak in which they are forcefully enrolled.

Where civets are represented in kopi luwak advertisements, they are depicted as sampling coffee cherries as if voluntarily, where they are situated amid greenery rather than within the confines of captivity. Alternatively, they are shown without surrounding context, the outline of a cat-like figure next to a coffee bean and a coffee cup that simply situates them not as living beings but as mere objects within the production process. In the case of tourism, civet statues stand strong and powerful, giving the impression of purpose and intent to their role in Balinese economies. It is therefore a rather fundamental oversight that civet–tourist encounters within kopi luwak tours lead tourists to question the ethics and legitimacy of the product's apparent heritage value.

Civet Exploitation as an Unethical Element of Heritage

Since the industry's emergence in 2010, kopi luwak tourism has enrolled civets into service in two fundamental ways. First, and most common, civets are housed in small cages as display items. Typically constructed from either wood or metal with wire mesh walls and a slatted floor (for ease of faecal collection), civet housing is inadequate to meet basic animal welfare needs such as shelter from the sun (of particular importance for a nocturnal animal). Due to the poor conditions endured, it was not uncommon for tourists to note instances of civet injury and stereotypic behaviour, conditions similarly reported by Carder *et al.*) in 2016. In partaking in activities such as pacing and self-mutilation, abnormal repetitive behaviours commonly associated with stress (Mason, 1991), civets made their 'resistance to captivity' obvious (Guenther,

Fig. 4.4. Left: A statue of a teeth-baring palm civet greets visitors at the entrance of a kopi luwak tour. Right: A two-story high civet statue. (Anonymous TripAdvisor reviewers.)

2020, p. 193). Such clear indicators of cruelty were recognized by tourists, some of whom used TripAdvisor to air their concerns and call for the boycott of kopi luwak tourism facilities, as did the following reviewer:

> It's interesting to see all the spice, fruit, and coffee trees, but then you come to the civet cats. These poor creatures are kept in bare cages and fed coffee beans to make wildly expensive coffee. They are nocturnal and need somewhere dark to sleep during the day. Instead, they were madly pacing up and down on bare boards. It was depressing to see. Don't contribute to this industry. Avoid the place.

In 2017, however, and likely in response to negative reviews concerning abuses of animals, new forms of bodily control began to emerge. The instances of the term 'cage' in tourist reviews began to become less common, and selfies with civets became popular features across most of the surveyed plantations. Unfortunately, both tourist reviews and uploaded photographs to the TripAdvisor site indicated that civets may be being drugged for safe handling by tourists, who were assured the animals were merely sleeping (Hooper, 2022b). The civet's nocturnal nature has therefore become another exploitable feature, provided as justification for the animal's sedentary appearance. One reviewer explained:

> Wild animals kept in cages, and one of them put on a table for people to pet it and take photos with it. Most likely it was drugged as it didn't move at all. I asked why and they just said that it is awake at night. The four ones in cages were oddly enough very active during the daytime.

Already stripped of their ability to select their own food and express species-typical behaviour, the liberation from a cage only to then be chemically immobilized is a form of concealed violence, one that completely debilitates the animal's agency. Some reviewers expressed their concerns about suspected abuse with a call to action on the TripAdvisor platform:

> Would not recommend this place solely on the fact that they had a luwak which was tranquilized at the shop for people to touch. Was just lying there like he was dead.

However, many tourists found the presence of 'sleeping' civets as another opportunity for consumption. Several reviewers claimed close contact interactions were the 'highlight' of the trip, as they were able to 'cuddle', 'hold' and 'pet' a civet. Thus, the sedation of civets is a clever ploy to bring tourists closer to nature while concealing the cruelty that led to negative reviews on the world's largest tourist review platform. Indeed, animal welfare concerns were most frequently associated with negative Trip Advisor reviews, and it was the barren metal cages which were the trigger.

For many tourists, the cage detracted from the fantasized imaginary of pristine nature, a pristine nature that modern zoos in economically advanced nations promote to their visitors and Bali tour operators promise to holidaymakers. For others, the narrative of wild animals sleeping was a welcome addition to the story of traditional mutualism between two separate species. The sleeping civet, apparently unconcerned by the disturbance of tourists, appeared to prove the phenomenon that civets would voluntarily partake in kopi luwak tourism and faecal commodification. Yet, the presence of cages did not disappear from kopi luwak agrotourism, as caged civets remained an integral part of the bean-to-cup production process displayed to tourists.

The Breakdown of Traditional Narratives: The Presence of Live Civets

The traditional story of kopi luwak, whereby local farmers collect digested coffee beans from the deposits of wild civets, is a story of 'speculative accumulation', wherein it is the mere appearance of value which instigates its material worth (Tsing, 2014, p. 75). While the reliance on wild collection affirms the product's rarity, a feature which ensures its market value and consumer demand, the profit is ensured by the claim of rarity, not the truth of it. Thus, what is observed in kopi luwak agrotourism is the undoing of the traditional narrative, the undoing of speculative accumulation despite wild collection being consistently promoted across the surveyed plantations.

First, tour guides recited the story of traditional wild collection throughout interactions with tourists during the guided tours. Second, the curated experience was consistently designed to meet the cultural heritage expectations of visitors. This was expressed not only in the tour design, which included the staging of plants, animals and humans as actors and participants in the kopi luwak production process, but also by the branding of human staff as responsible parties in wild faecal collection. In one site, staff were identified by t-shirts with the slogan 'Poo Hunters' branded across the chest. Video recordings and written reviews substantiated that, when questioned, tour guides would confirm they collect the faeces from the forest, despite no evidence to confirm the presence of locally free-roaming civets. One such reviewer noted their concerns by highlighting the inconsistency in the claims of tour guides versus their own observations:

> The tour guide told me the civets are released each night and then put back in the cages each morning ... but when I observed the locks they were covered in cobwebs and rusty wire. I don't know if he was telling the truth.

While the story of kopi luwak heritage was consistently promoted across sites, the authenticity of the traditional narrative was undone in several ways. Primarily, the large quantity of kopi luwak available for tasting and purchase in each site combined with the high number of plantations operating along the tourist trail undermines the claims of global rarity. The false claims of a global availability of only 127 kg per year is a central facet on which kopi luwak's exclusive status relies. Indeed, the low number of displayed civets led many tourists to question the authenticity of such sites. Many noted that the quantity of kopi luwak available for purchase in the gift shop far exceeded the display civet's capacity for production. One reviewer wrote: 'How do they get so much coffee from the 12 or so civets on display? ... is there a "Luwak factory" somewhere where they produce the digested beans, akin to caged egg farming?'

Interestingly, where questions of authenticity were raised through the presence of captive civets, such answers remained uncertain. However, reviewers who were led to question the touristic authenticity through the purchase of fake products were much more certain in their accusations towards the industry of false product claims.

The Breakdown of Trust between Tour Operators and Tourists: The Sale of Fake Products

Reviewers frequently used the term 'tourist trap' to purvey the dishonest capture of tourist revenue through illegitimate enterprises, a term that dehumanizes the tour operator (as the predator) and the tourist (as the prey). Many tourists noted the number of similar products available much more cheaply in local supermarkets, and one reviewer even claimed to see a delivery of coffee bags waiting to be re-packaged and put on display in the shop.

Most prevalent amid the reviews ranked as 'terrible' were complaints due to sale of fake products, including tea, coffee and kopi luwak. Hundreds of reviewers described similar scenes, having spent hundreds of dollars each on '100% organic' teas, coffees, chocolates and wine, only to find their purchases weighed less than stated on the packet and contained nothing more than powdered sugar, flavourings or instant coffee. The realization that faulty produce has been sold at, as one reviewer described, 'wildly inflated prices', degraded the entire memory of the experience, with reviewers claiming they had been 'lied to', 'scammed' and 'conned'. For many, fake products cast the entire experience into doubt – the disillusionment shattered the natureculture premise and impacted their sense of self as supporters of native communities and ecologies. The feeling of being cheated revealed an orientalist vitriol whereby the tourist saw themselves as the saviour of traditional native practices. One such reviewer explained:

> Since they try to sell visitors poor quality or flat out non-genuine products, I highly doubt Sarita is run by an actual 'farming family' as they tried to represent themselves as and that my tourist dollars actually went to help some local farmers.

For some reviewers it was only once the legitimacy of the product had been cast into doubt that their attentions turned towards the treatment of the display animals, as illustrated by one reviewer who concluded a very lengthy review of faulty products with the following reflection:

> Yes, the cats are kept in pretty sad conditions, I'm sorry I bought anything from there.

In such cases, animal welfare was a secondary concern and Pachirat's (2011) caution as to the politics of sight were confirmed. Simply witnessing civet abuse was not enough to deter tourists from financially supporting the industry through purchasing of goods. Thus, through the perspective of posthumanism, it is clear to see that despite the allure of nature enticing tourists to see plants and animals, kopi luwak tourism is inherently anthropocentric. In all, kopi luwak tourism tends to prioritize the economic interests of humans over the intrinsic interests of animals. It was only when purchased goods were found to be illegitimate that civet welfare was of ethical concern and support for the local business was withdrawn. The secretive nature of the production line combined with the difficulty to distinguish wild-collected from caged-produced kopi luwak beans (Jumhawan *et al.*, 2013) remain problematic for ethical consumerism. The curated experience within kopi luwak agrotourism simply highlights this problem. The display of captive civets combined with the telling of untruths by tour guides presents a paradoxical narrative that undoes the traditional storyline on which the kopi luwak industry is built, revealing instead the previously concealed ethical implications of mass caged production.

Conclusions

In all, TripAdvisor offered an insight into tourist perceptions of Balinese kopi luwak agrotourism sites since their emergence in the early 2000s. However, despite virtual ethnography offering a rich dataset to anthropological practice (O'Connor, 2010), it is important to note that this chapter cannot give voice to the tour operators. According to the 18 Balinese ethical moral values, 'the attitude of being honest' ranks as third importance after 'the religious attitude' and 'the attitude of being tolerant' (Suwija *et al.*, 2019). Thus, enrolment in an industry that relies on a now false traditional narrative, whereby tourists are purposefully misled, could in fact cause psychological harm to the staff employed there. Furthermore, the branding of staff as 'Poo Hunters' and the placement of elderly women within the kitchen reinforces the colonial and gender hierarchies which perpetuate inequality. Particularly troubling were the colonial narratives prevalent among tourists seeking to immerse themselves in the traditional practices of an industry that is built on the othering of indigenous humans and animals, for it was the forbidding of farmers to consume their own crops that led to kopi luwak's origins. It would be highly advantageous for further research into the kopi luwak tourism industry to involve participant interviews with staff and immersive longitudinal ethnographic investigation.

In all, a posthumanist investigation of kopi luwak tourism has highlighted the overlooked importance of multi-species encounter within emergent tourism practices. The presence of animals as out of place transforms the story from one of genuine cultural tradition and rare commodity to one of suffering, exploitation and control. In short, kopi luwak tourism exposes the corruption of the tradition, revealing the truth of contemporary venture whereby plants, humans and animals are enmeshed in the capitalization of a fetishized experience. Kopi luwak agrotourism is inauthentic, an enterprise based on the tourist's idea of cultural heritage and pristine nature.

References

Adams, C. (2018) *The Sexual Politics of Meat*. Routledge, Oxford, UK. DOI: 10.4324/9780429499142.

Baudrillard, J. (1981) *Selected Writings: Simulacra and Simulation*. Polity Press, Cambridge, UK.

Belhassen, Y. and Caton, K. (2006) Authenticity matters. *Annals of Tourism Research* 33(3), 853–856. DOI: 10.1016/j.annals.2006.03.009.

Bruncevic, M. and Linne, P.A. (2018) *A Taste of Law and Coffee-From Tastescape to Lawscape*. Law and the Senses Westminster Law & Theory Lab Series, p. 201. DOI: 10.16997/book21.

Buchmann, A., Moore, K. and Fisher, D. (2010) Experiencing film tourism: authenticity and fellowship. *Annals of Tourism Research* 37(1), 229–248.

Cahill, C. (2020) [Un]becoming a resource: translating the nature of civets in Indonesia. *Ethnos* 85(1), 100–117. DOI: 10.1080/00141844.2018.1536072.

Carder, G., Proctor, H., Schmidt-Burbach, J. and D'Cruze, N. (2016) The animal welfare implications of civet coffee tourism in Bali. *Animal Welfare* 25(2), 199–205. DOI: 10.7120/09627286.25.2.199.

Cohen, E. (2019) Posthumanism and tourism. *Tourism Review* 74(3), 416–427. DOI: 10.1108/TR-06-2018-0089.

Cole, S. (2012) A political ecology of water equity and tourism: a case study from Bali. *Annals of Tourism Research* 39(2), 1221–1241. DOI: 10.1016/j.annals.2012.01.003.

Collins-Kreiner, N. and Wall, G. (2015) Tourism and religion: spiritual journeys and their consequences. In: Brunn, S.D. (ed.) *The Changing World Religion Map*. Springer, New York, pp. 689–707. DOI: 10.1007/978-94-017-9376-6.

Douglas, M. (2003) *Purity and Danger: An Analysis of Concepts of Pollution and Taboo*. Routledge, Oxford, UK. DOI: 10.4324/9780203361832.

Everett, S. (2008) Beyond the visual gaze? The pursuit of an embodied experience through food tourism. *Tourist Studies* 8(3), 337–358. DOI: 10.1177/1468797608100594.

Fuentes, A. (2010) Naturalcultural encounters in Bali: monkeys, temples, tourists, and ethnoprimatology. *Cultural Anthropology* 25(4), 600–624. DOI: 10.1111/j.1548-1360.2010.01071.x.

Glaesser, D., Kester, J., Paulose, H., Alizadeh, A. and Valentin, B. (2017) Global travel patterns: an overview. *Journal of Travel Medicine* 24(4). DOI: 10.1093/jtm/tax007.

Guenther, K.M. (2020) *The Lives and Deaths of Shelter Animals: The Lives and Deaths of Shelter Animals*. Stanford University Press, Stanford, California. DOI: 10.1515/9781503612860.

Hakim, L., Siswanto, D., Rahardi, B. and Zayadi, H. (2019) Fostering coffee agroforestry for agrotourism development in degraded land in a buffer zone of a national park: a case study from Poncokusumo, Malang, Indonesia. *EurAsian Journal of BioSciences* 13(2), 1613–1620.

Haraway, D.J. (2003) *The Companion Species Manifesto: Dogs, People, and Significant Otherness*, Vol. 1. Prickly Paradigm Press, Chicago, Illinois.

Haraway, D.J. (2008) *When Species Meet*. University of Minnesota Press, Minneapolis, Minnesota.
Hooper, J. (2022a) Cat-Poo-Chino and captive wildlife: tourist perceptions of Balinese kopi luwak agrotourism. *Society & Animals* 1–20. DOI: 10.1163/15685306-bja10094.
Hooper, J. (2022b) Contamination: the case of civets, companionship, COVID, and SARS. *Journal of Applied Animal Welfare Science* 25(2), 167–179. DOI: 10.1080/10888705.2022.2028627.
Jumhawan, U., Putri, S.P., Marwani, E., Bamba, T. *et al.* (2013) Selection of discriminant markers for authentication of Asian palm civet coffee (kopi luwak): a metabolomics approach. *Journal of Agricultural and Food Chemistry* 61(33), 7994–8001. DOI: 10.1021/jf401819s.
Kellert, S.R. and Wilson, E.O. (1993) *The Biophilia Hypothesis*. Island Press, Washington.
Lachenmeier, D.W. and Schwarz, S. (2021) Digested civet coffee beans (kopi luwak) – an unfortunate trend in specialty coffee caused by mislabeling of *Coffea liberica*? *Foods* 10(6), 1–4. DOI: 10.3390/foods10061329.
Lovell, J. and Hitchmough, S. (2020) Simulated authenticity: storytelling and mythic space on the hyper-frontier in Buffalo Bill's wild west and westworld. *Tourist Studies* 20(4), 409–428. DOI: 10.1177/1468797620937912.
Marcone, M.F. (2004) Composition and properties of Indonesian palm civet coffee (kopi luwak) and Ethiopian civet coffee. *Food Research International* 37(9), 901–912. DOI: 10.1016/j.foodres.2004.05.008.
Mason, G.J. (1991) Stereotypies: a critical review. *Animal Behaviour* 41(6), 1015–1037. DOI: 10.1016/S0003-3472(05)80640-2.
Muzaifa, M., Hasni, D. and Rahmi, F. (2019) What is kopi luwak? A literature review on production, quality and problems. In: *Earth and Environmental Science*, Vol. 365. IOP Publishing, Bristol, UK, p. 012041. DOI: 10.1088/1755-1315/365/1/012041.
Muzaifa, M., Hasni, D., Patria, A. and Abubakar, A. (2018) Sensory and microbial characteristics of civet coffee. *International Journal on Advanced Science, Engineering and Information Technology* 8(1), 165. DOI: 10.18517/ijaseit.8.1.3092.
Nekaris, B.K.A-I., Campbell, N., Coggins, T.G., Rode, E.J. and Nijman, V. (2013) Tickled to death: analysing public perceptions of 'cute' videos of threatened species (slow lorises – *Nycticebus* spp.) on web 2.0 sites. *PloS One* 8(7), e69215. DOI: 10.1371/journal.pone.0069215.
Nicolaides, A. (2020) Sustainable ethical tourism (SET) and rural community involvement. *African Journal of Hospitality, Tourism and Leisure* 9(1), 1–16.
O'Connor, P. (2010) Managing a hotel's image on TripAdvisor. *Journal of Hospitality Marketing & Management* 19(7), 754–772. DOI: 10.1080/19368623.2010.508007.
OECD Tourism Statistics (2022) Indonesia: tourism in the economy. OECD Tourism Statistics. DOI: 10.1787/888934078547
Pachirat, T. (2011) *Every Twelve Seconds: Industrialized Slaughter and the Politics of Sight*. Yale University Press, Yale, Connecticut.
Reisinger, Y. and Steiner, C. (2006) Reconceptualising interpretation: the role of tour guides in authentic tourism. *Current Issues in Tourism* 9(6), 481–498. DOI: 10.2167/cit280.0.
Smith, S.P. (2018) Instagram abroad: performance, consumption and colonial narrative in tourism. *Postcolonial Studies* 21(2), 172–191. DOI: 10.1080/13688790.2018.1461173.
Straits Research (2022) Kopi luwak coffee market size is projected to reach USD 10 billion by 2030, growing at a CAGR of 4.9%: straits research. Available at: https://www.globenewswire.com/en/news-release/2022/10/04/2528177/0/en/Kopi-Luwak-Coffee-Market-Size-is-projected-to-reach-USD-10-Billion-by-2030-growing-at-a-CAGR-of-4-9-Straits-Research.html (accessed 23 November 2022).
Suwija, I.N., Suarta, I.M., Suparsa, I.N., Alit Geria, A.A.G. and Suryasa, W. (2019) Balinese speech system towards speaker social behavior. *Humanities & Social Sciences Reviews* 7(5), 32–40. DOI: 10.18510/hssr.2019.754.
Tsing, A.L. (1993) *In the Realm of the Diamond Queen: Marginality in an out-of-the-Way Place*. Princeton University Press, Princeton, New Jersey. DOI: 10.1515/9781400843473.
Tsing, A.L. (2014) Blasted landscapes (and the gentle arts of mushroom picking). *The Multispecies Salon* 87–109. DOI: 10.1215/9780822376989.
Tucker, C.M. (2017) *Coffee Culture: Local Experiences, Global Connections*. Taylor & Francis, Oxford, UK. DOI: 10.4324/9781315678795.
UN (2018) Revision of the World Urbanization Prospects. Available at: https://population.un.org/wup/ (accessed 26 July 2023).
Urry, J. (2002) *The Tourist Gaze*. Sage, London.

5 Animal Trade and Dark Tourism in Sonora Market

Jean Azcatl Pineda* **and Alicia Mariana Penélope Castro Pérez**
Universidad Nacional Autónoma de México, Mexico City, Mexico

Abstract

Sonora Market is an important traditional commercial space located in the centre of Mexico City. It offers, among other items, supplies for practising witchcraft and Santería rituals, making it a mystical and distinctive place for those seeking to connect with Mexican 'otherness'. However, live animals are also sold there. They are used in these practices – considered intangible cultural heritage – and their presence is crucial to the tourist experience, as they reinforce the imagery of this space being a traditional marketplace. In this chapter, built on mixed methods, it is debated why this market is an 'anti-place', and why the sale of animals, based on speciesist violence that transgresses their rights and interests, is unethical.

Introduction

Traditional markets in Mexico play a fundamental social role. Many of the local culture habits and traditions are expressed in their constant flow of goods and commodities (Anzures, 1991; Bellucci, 2002). At Sonora Market, one of Mexico City's largest, it is possible to find anything from handicrafts and toys to items used in Santería, such as esoteric ritual objects and traditional herbal medicine. Live animals, too, are traded for their ornamental qualities and as ingredients in the preparation of amulets and remedies.

Its peculiar commercial dynamics and its proximity to Mexico City's historic centre make this market a popular tourist attraction for many national and international visitors interested in popular lore, traditional medicine and magic. Animals, however, live in miserable confinement and can lose their life at any time (Martínez, 2010; Barquero, 2017). This situation makes Sonora Market a place of insecurity and danger for most animals, an anti-place (Méndez, 2012; Godoy and Rodríguez, 2020) in which they are forced into captivity, deprived of their autonomy and agency, and treated solely as a commodity.

The aim of this chapter is twofold: (i) to submit the practices of captivity and commodification of animals to a deontological analysis from a posthumanist anti-speciesist perspective (Singer, 1975; Cooney, 2003; Smith, 2003; Francesca, 2019; López and Quintero, 2021b) and (ii) to frame these practices as dark tourism in the context of dissonant heritage (Tunbridge and Ashworth, 1996).

Research for this chapter was conducted with a combined methodology that included: (i) bibliographic searches on animal geography, tourism geography and animal ethics; (ii) 'web-scraping' of social media platforms for reports by visitors to the market (Marres and Weltevrede, 2013; Rodríguez, 2022); (iii) web analysis of data obtained from TripAdvisor;

*Corresponding author: jean.geohum@gmail.com

 Tourism, Heritage and Commodification of Non-human Animals (eds A. López López *et al.*)
DOI: 10.1079/9781800623309.0005

(iv) fieldwork in Sonora Market between May and December 2021 based on the participant-observation technique, in order to observe the commercial dynamics of animal trade (López and Van Broeck, 2013); (v) semi-structured interviews to explore visitors' perceptions of the situation of animals in the market; and (vi) the compilation of a photographic registry on the conditions in which the animals live. First, a theoretical approach to tourism geography involving animals and dark tourism in places of dissonant heritage is needed.

Tourism as the Consumption of Otherness

Tourism is a global and globalising activity bringing exogenous agents and local communities together (Portilla, 2002; Aramberri, 2008; González and Castañeda, 2012). As exogenous agents, tourists are collectors of extraordinary experiences that allow them to situate themselves in the world and in society (Gibson, 2012). In Mexico, tourism has been seen as a tool for economic reactivation, job creation and regional development (Espinoza *et al.*, 2012). However, the depredatory side of tourism has also been exposed, as has its socio-economic impacts and its role in the promotion of an anthropocentric ideology that reifies animals (Brown and Hall, 2008; Chávez, 2019; Quintero and Castro, 2022).

The driving forces of cultural tourism are the desire to collect new experiences, the search for authenticity and the consumption of the 'Other' (Ateljevic and Doorne, 2008). The perception of otherness by the observer depends on several factors, such as value system, gender, economic status, geographical origin and education (Urry and Larsen, 2011; Chávez, 2019). In a process that lends to Foucauldian analysis, otherness reifies individuals, societies and their practices and territories, making them amenable to consumption as resources and commodities (Sharpley, 2022). It is important to note that tourism is not only about seeing, and that the encounters with otherness are multi-sensorial, that is, they involve the other human senses (Gibson, 2012).

Through differentiation and recognition, locations with a palpable cultural heritage compete fiercely for a position in the cultural tourism market (Foroudi *et al.*, 2019; Tena, 2020). Within cities, the spaces valued as 'authentic' for the preservation in them of local (viz. 'exotic') practices are consumed (Gibson, 2012; Crespi and Domínguez, 2016), and animals are often involved.

Local markets are open to tourist consumption because they are thought to possess an authentic otherness, expressed in their specialty products (Crespi and Domínguez, 2016; García, 2017). In Mexico City, such markets not only satisfy local and regional demand for goods and services but also have been turned into tourist attractions that facilitate the encounter between locals and newcomers. Cultural practices thus become identity referents and are transformed into intangible heritage (Everett, 2008; Crespi and Domínguez, 2016; García, 2017).

Usually deprived of their primary interests and fundamental rights, animals are omnipresent in the tourism system (Mkono and Holder, 2019; Conanp, 2020). The Universal Declaration on Animal Rights promulgates the rights of all beings to lead a life free from hunger and thirst, sickness, pain or injury, fear and anxiety, and to thrive according to their own behavioural patterns (Neumann, 2012). Under this premise, any form of tourism that reifies and exploits animals and benefits from their suffering and death transgresses these basic tenets.

In traditional Mexican markets, animal carcasses are consumed for culinary, medicinal or magical purposes, but live animals are also traded, and they lead a life of confinement, crammed together for long periods (Ortiz Millán, 2020; Azcatl, 2021). Such practices become customs and in time end up being institutionalized as heritage (Prats, 1998; Macdonald, 2009). The naming of heritage involves a selection of a few practices in exclusion of others, which leads to disagreement, especially when the death and suffering of animals are involved (Tunbridge and Ashworth, 1996). Heritage that includes practices associated with death, disaster, conflict or trauma and over which there is disagreement is known as atrocity heritage (Tunbridge and Ashworth, 1996; Macdonald, 2009), difficult heritage (Castro and Wójtowicz, 2018) or negative heritage (Rico, 2008). Heritage can be more or less dissonant according to the type of atrocity, the nature of cruelty and the perpetrators, the visibility of the event and the permanence of the registry (Tunbridge and Ashworth, 1996). According to this classification, heritage that includes cultural practices involving the exploitation, suffering and death of animals ought to be classified in the most difficult and dissonant categories, given the violence involved against vulnerable subjects in a systematic and sustained way.

Similarly, dark tourism refers to places where suffering and/or death are exploited as tourist attractions (López and Quintero, 2021a). Originally a category for the analysis of human practices (Lennon and Foley, 2000), dark tourism was introduced by Quintero and López (2017; López and Quintero, 2021a) to analyse cultural tourism practices involving the suffering and death of animals and legitimized by tradition and heritage.

Tourism and Consumption of Animals in Sonora Market

The proximity of Sonora Market to Mexico City's historic centre (Fig. 5.1) favours not only the flow of goods but also of visitors (Azcatl, 2021), who attribute cultural symbols to it (Tuan, 1994; Hiernaux and Lindón, 2006). The market is organized in four main sections: handicrafts and ceramics, toys and disguises, herbalism and esotericism, and live animals.

Tourists feel surrounded by an atmosphere of sensations on account of the many traditions it contains. Our fieldwork registered visitors from many countries, including, in order of statistical importance, the USA, Brazil, Japan, Spain, Peru, Costa Rica, Australia, South Africa, New Zealand, Canada, Turkey, China, Poland and Russia. These people perceived the market as an authentic Mexican space.

The analysis of the tourist imaginaries of Sonora Market relied on the information contained on tripadvisor.com.mx (accessed January 2023).

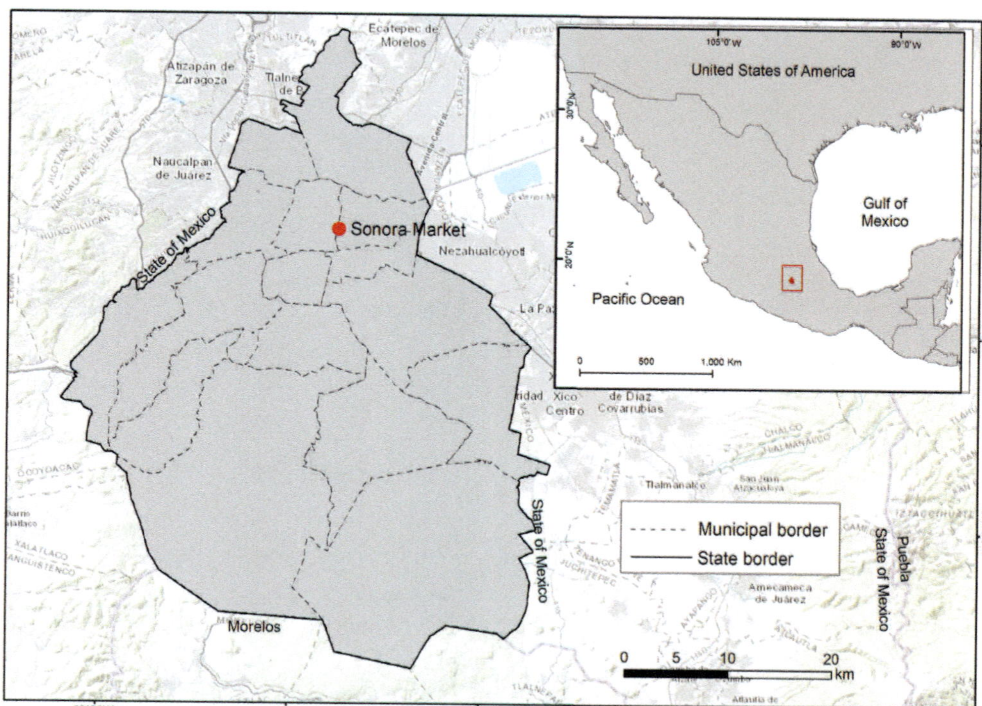

Fig. 5.1. Location of Sonora Market in Mexico City.

This comprehensive tourist information site features a special section in which visitors can freely write their opinions and evaluations and share their experiences during their visit. The statistical sample consisted of 68 visitor reviews on Sonora Market written until 1 January 2023. The information was classified by nationality and gender.

Visitors writing on this section of TripAdvisor can rate the destination in question using a system of stars that goes from 1 to 5. In some instances, they state their nationality or the place they come from. To process the information, the software ScrapeHero was used. The information was classified according to: (i) the main areas of interest of the destination; (ii) the presence or absence of animals in the experiences; (iii) the assessment, from the tourists' perspective, of the conditions in which these animals are kept; and (iv) the uses to which they are put in the different sections of the market.

After the information was processed, two webs were built with Gephi software, one on the visitors' general perceptions of Sonora Market and another one on their perceptions of the condition of animals. The distribution algorithm was ForceAtlas2, which allows to visualize the node communities with greater clarity. The concepts expressed by visitors appear as nodes in the web and their size varies in relation to the frequency with which they appeared on the site. Also, a node's position indicates its relative centrality in terms of the market's general perception.

The analysis of this data led to the conclusion that the main areas of interest for tourists inside the market were esotericism, witchcraft and live animals (Fig. 5.2). Forty-six per cent of

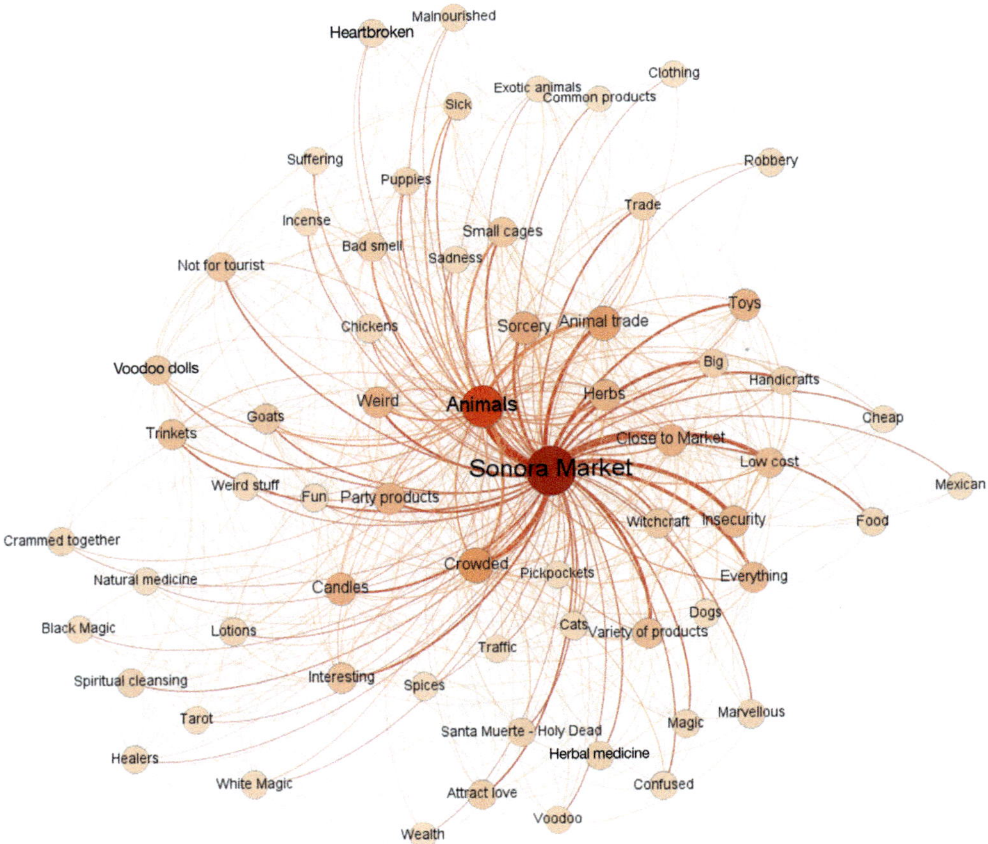

Fig. 5.2. Concepts found in visitors' reviews of Sonora Market on TripAdvisor. (Based on TripAdvisor information, 2022.)

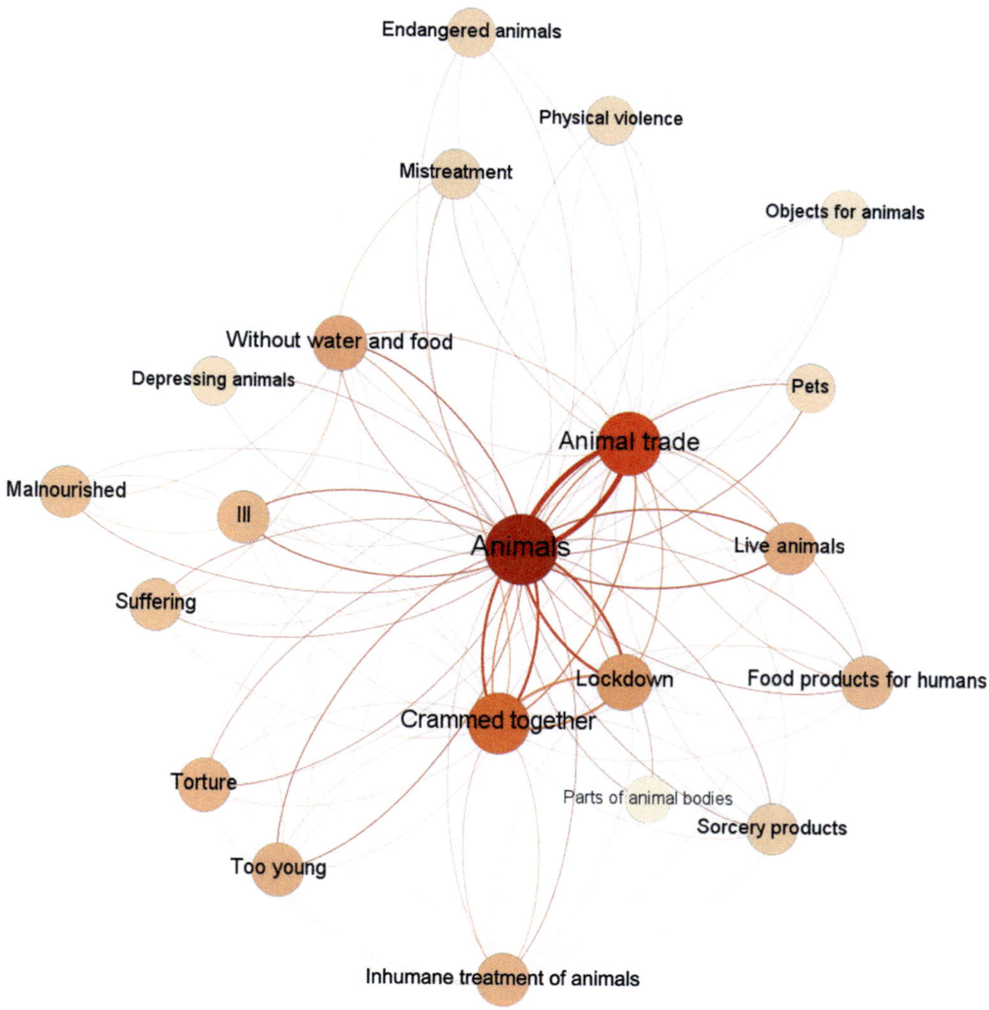

Fig. 5.3. Animals in Sonora Market. (Based on TripAdvisor information, 2022.)

the comments had to do with animals (mostly live ones), which occupy a central position in the tourist experience. Theirs is the second largest node and it connects with the first one (Fig. 5.2). The node of animals is the origin of a group of concepts associated with the diversity of species present in the market, the problems caused by confinement and the practices in which they are used, such as esotericism and Santería.

Fig. 5.3 shows in more detail the problems visitors identify in the animals' living conditions, such as confinement, cramming, deprivation of food and water, and the presence of endangered species and of young individuals. Most of the visitors commenting on TripAdvisor reported having seen torture, physical violence, abuse, malnutrition, sickness and suffering.

Sonora Market was described in these reviews as a colourful place featuring overflowing fruit and vegetable stalls and price tags in a noisy atmosphere marked by an incessant movement of people, a wide variety of scents and other sensations. The sounds that impregnate the environment include loud mongering for miracle remedies for any ailment, made

Fig. 5.4. Sale of live birds at Sonora Market. (Fieldwork photo.)

from axolotl – an endangered Mexican amphibian (Encyclopedia of Life, 2023) – paste and every other imaginable animal ingredients. Live or dead animals used in witchcraft (such as canaries, hummingbirds, all sorts of reptiles and even young goats), alternative medicine (axolotls, armadillos, skunks and several bird parts) or food (cows, birds, swine) are perceived as part of cultural heritage, and their exploitation remains imperceptible, absent referents (Adams, 2016).

Though the market sections devoted to Santería sell animal products and not live animals, their odours blend with those of amulets, totems and brews made with wild animal parts or extracts, and other magical-religious objects. Rabbit and monkey feet, deer antlers, puma paws, exotic bird feathers, armadillo shells, skins and even whole dead animals are sold with no regulation whatsoever (Martínez, 2010).

Of the four sections of the market, the one devoted to live animals is deemed by visitors as the one that most engages the senses. Screeches, howling and chirping compete with the mongers' voices. Victims of this trade are an endless number of birds, goats, dogs, cats, reptiles (including small alligators), insects, rodents and other reified creatures (Figs 5.4 and 5.5). These animals urinate and defecate in

Fig. 5.5. Sale of live puppies at Sonora Market. (Fieldwork photo.)

their cages, and despite the constant cleaning of aisles, the stench is hardly eliminated and combines with that of the market's garbage dump.

According to the De la Cruz-Badiano Codex (Linares and Bye, 2013; Fernández, 2014), remedies with animal ingredients were common among pre-Hispanic peoples. Many of these can still be found today in many markets of the Valley of Mexico, and some are used in witchcraft (Gómez *et al.*, 2007). Since their effectiveness has no scientific backing, the sacrifice of the animals to produce them remains a frivolous practice that contributes, among other things, to the pillaging of ecosystems and the displacement and loss of wild fauna (Anzures, 1991; Sarukhán, 2009).

Sonora Market is perceived as an 'authentic' place on account of the presence in it of exotic objects, live animals and traditions regarded as pre-Hispanic heritage. Opinions go from the exalting to the morbid:

> A market where you need to be extra careful, have to be prepared, not to take important things ... Just be aware, not safe, pickpockets, not very clean and safe, the real third world, if you can handle that, and feel adventurous, it is worthy. Prices are the best, can ask for discounts, many rare things, a real colourful, real market, second in the world for black magic, if you know what I mean, can get lots different things, from food, candles saints too many things to describe. A unique, stuff you don't get at touristic markets.
>
> (Tourist, USA, 2015; TripAdvisor, 2022)

> This place isn't for everyone. If you are sensitive to the inhumane treatment of animals, cramped conditions, claustrophobia, etc. this is NOT the place for you. However, I think travel is about experiencing and seeing things that expand your worldview and give you an experience outside of what you are comfortable with. I was aware of the inhumane treatment of animals before visiting, but I checked out that section out of morbid curiosity. It is definitely sad, intense, and could be traumatizing for young people. I am also glad I saw it ... You have to go to the back to find the really weird stuff ... Once you make it to the back you will be rewarded with beautiful paper mâché masks, religious objects, very strangely designed packaging, spices, and a variety of other objects and ephemera I guarantee you won't find anywhere else. To me this is a must visit for CDMX. You will be uncomfortable, but that is the fun.
>
> (Tourist, unknown origin, 2019; TripAdvisor, 2022)

Tourist imaginaries range from the indifference of the privileged tourist (Hollinshead, 1999) who, disengaged from animal suffering, sees the market as exciting for its dangerous, wretched and backward quality, to the awareness of those who criticize its *modus operandi* – though always from a dominant anthropocentric stand. Thus, the dark experiences with animals exist both in their tangible exploitation and in the imagination of tourists.

Sonora Market as an Anti-place: Ethical Issues

Speciesism is an ideology that privileges human interests over those of other beings, and a moral discrimination that has perpetuated exploitative practices and the enslavement and extermination of animals (Singer, 1975). Sonora Market is the epitome of speciesism because there animals are treated as objects and not as sentient beings with their own needs and desires (Mata, 2020). While visitors experience the market as a relatively safe 'place', the animals experience it as an anti-place of fear, violence and insecurity (Godoy and Rodríguez, 2020).

Several perspectives on the quality of place have been developed. There is the place of belonging and security (Tuan, 1974), the no-place of anonymity and clandestinity (Augé, 1992), and the anti-place of suffering, domination and systematic death (Méndez, 2012). A place may have negative connotations for individuals, who then experience *topophobia*, that is, a subjective aversion to specific spaces (Tuan, 1974). But the anti-place is a landscape of collective rejection inasmuch as it is linked to crime and violence (Méndez, 2012). The anti-place also generates its own imaginaries, charged with explicit violence, real and perpetual fear of abuse, of torture, confinement and hunger, disease, pain and death, as happens to the animals in Sonora Market (Fig. 5.6). Though the concept of anti-place was developed to describe the imaginaries

Fig. 5.6. Birds in crates. (Fieldwork photo.)

of human fear (Godoy and Rodríguez, 2020), it is perfectly relevant to the case of animals, who are sentient beings conscious of their suffering (Low, 2012). As Tuan (1976) points out, spatial awareness is an animal instinct, and this applies to the animals in Sonora Market.

Animal rights groups have reported the systematic violence inflicted on animals in Mexico City. On the other hand, the psychosis generated by the fear of zoonosis as a result of the COVID-19 pandemic (Azcatl and Quintero, 2022) has led to the prohibition of the sale of animals in street markets (Mexico City's Government Gazette, 2020, p. 14). Thus, Sonora Market is not only an anti-place for animals in confinement but also a space of illegal animal violence and slaughter by organized groups. Regardless of their perception of it as a place, a no place, a place of rejection or dissonance, the fact remains that tourists do not refrain from wanting to 'live' the experience.

Sonora Market is a dark-tourism space because of the imagined community of visitors associated with it. Tourists are keen to come close to heritage practices involving tragedy and suffering (Frew and White, 2013). The presence of tourists in Sonora Market legitimizes and reaffirms the suffering of animals from a hierarchical position that may or may not be voluntary but that eventually justifies the permanence of this anti-place.

The avoidance of suffering is a primary interest of all sentient beings. Apart from guaranteeing them access to food and water, animals should be free from pain, anxiety and fear, and the kind of injuries inflicted on them (Brambell, 1965) that are common at Sonora Market. The Global Code of Ethics for Tourism (UNWTO, 2023) has yet to include strict rules on the participation of animals – hitherto regarded simply as resources – in tourism activities, given the varying degrees of injustice to which they are subjected and that they are deprived of their autonomy, dignity and respect (Fennell and Sheppard, 2020). From a posthumanist anti-speciesist perspective, tourism remains an economic force that contributes to the oppression of other species.

In Mexico, the conservation of animal species is sought in the General Law of Wildlife, Conservation and Sustainable Use (Semarnat, 2017), and the Mexico City's Government Gazette (2020) establishes the prohibition of the sale and exploitation of animals. However, these practices continue to go unpunished. As long as the laws remain unenforced, the only relevant agency that remains is tourists' ethical decisions to support or boycott them. Animal tourism is not an indispensable activity for human survival, and support for these kinds of practices cannot be justified, even from the standpoint of utilitarianism (Singer, 1975) or abolitionism (Francione, 2010).

Conclusions

Sonora Market in Mexico City has been the locus of much debate due to the trade of live and dead animals, some of which are endangered species. Specially targeted have been the precarious conditions in which animals are kept, characterized by overcrowding and neglect. From a posthumanist perspective, these stressing conditions are tantamount to animal abuse. Furthermore, the market represents a public health hazard, due to the possibility of transmission of zoonotic diseases. The absence of control as to the origin of the species sold makes it reasonable to assume that the businesses are linked to the illegal trafficking of species, which has brought the attention of many animal rights groups.

However, from an anthropocentric perspective, the trade of animals in the market fulfils several purposes: the presumed conservation of cultural practices regarded as intangible heritage, the preservation of an image as a place of traditional trade and, of course, the economic benefits derived from the abuse and slaughter of animals. A space of confinement, violence and suffering is thus turned into an anti-place that attracts visitors.

The question arises as to how the conversion of such places of violence and suffering into tourist attractions is ethically viable: are the needs for recreation, culture/heritage preservation and economic gain enough reasons? The postulates of the Global Code of Ethics for Tourism are meant to induce tourists to respect the traditions, customs and practices of the places they visit. But what happens when these are anti-places of suffering propped up by illegal practices?

Tourists in search of 'original' and 'authentic' experiences should assume responsibility for the impact their actions have on the places they visit. They should be aware that the relative ambiguity of their status as exogenous agents often allows them to transgress social norms and ethical concerns at their travel destinations, and that in this case the act of consuming a space in which animal exploitation is rampant puts them in an ethical quagmire. On the other hand, the Global Code of Ethics for Tourism must be prompted to include animal rights in its guidelines. Beyond that, a wholesale deconstruction of collective thinking on the rights and freedoms of living beings in social, cultural and political terms is in order.

For years, tourism has thrived on the idea that selling experiences is a powerful economic engine. The designation of certain social and cultural practices as heritage goes in that direction. But when in the name of preserving cultural practices for future generations the exploitation, abuse and suffering of other living beings are legitimized, heritage becomes problematic. The search for profit has often neglected the ethical aspects of what is sold as 'authentic'. The existing legislation on animal rights must be expanded and enforced so as to guarantee that no tourism practice engages on any kind of exploitation and abuse. Until this happens, heritage dissonances will continue to exist.

References

Adams, C.J. (2016) *La Política Sexual de la Carne: una Teoría Crítica Feminista Vegetariana (The Sexual Politics of Meat: A Vegetarian Feminist Critical Theory)*. Ochodoscuatro Ediciones, Santiago, Chile.

Anzures, M. (1991) Etnología. El mercado de Sonora. *Anales de Antropología* 28(1), 273–290. DOI: 22201/iia.24486221e.1991.1.13095.

Aramberri, J. (2008) The future of tourism and globalization: some critical remarks. *Futures* 41(6), 367–376. DOI: 10.1016/j.futures.2008.11.015.

Ateljevic, I. and Doorne, S. (2008) Dialectics of authentication: performing 'exotic otherness' in a backpacker enclave of Dali, China. *Journal of Tourism and Cultural Change* 3(1), 1–17. DOI: 10.1080/14766820508669093.

Augé, M. (1992) *Non-Lieux: Introduction à Une Anthropologie de la Surmodernité*. Éditions Seuil, Paris.

Azcatl, J. (2021) Construcción espacial y valorización ética de las prácticas comerciales en el mercado de Sonora, CDMX desde una perspectiva de la geografía de los animales (Spatial construction and ethical assessment of commercial practices in the Sonora market, Mexico city, from an animal-geography perspective). Undergraduate thesis. National University of Mexico, Mexico City.

Azcatl, J. and Quintero, J. (2022) El mercado de Sonora, un arca de anacronismos en Ciudad de México (The Sonora market, an ark of anachronisms in Mexico city). *Animal Político*. Available at: https://www.animalpolitico.com/analisis/organizaciones/una-vida-examinada-reflexiones-bioeticas/el-mercado-de-sonora-un-arca-de-anacronismos-en-ciudad-de-mexico (accessed 12 January 2023).

Barquero, S. (2017) *Animales, la revolución pendiente*. La Esfera de los Libros, Madrid.

Bellucci, A.P. (2002) La herbolaria en los mercados tradicionales (Herbalism in traditional markets). *Revista Del Centro de Investigación de La Universidad La Salle* 5(17–18), 63.

Brambell, F. (1965) *Report of the Technical Committee to Enquire into the Welfare of Animals Kept Under Intensive Livestock Husbandry Systems*. Her Majesty's Stationary Office, London.

Brown, F. and Hall, D. (2008) Tourism and development in the global south: the issues. *Third World Quarterly* 29(5), 839–849. DOI: 10.1080/01436590802105967.

Castro, P. and Wójtowicz, M. (2018) Dibujando un sitio de atrocidad y tragedia: la percepción turística del "patrimonio difícil" del Museo de Auschwitz, Birkenau, Polonia (Picturing a site of atrocity and tragedy: the perception of the 'difficult heritage' of the Auschwitz Museum, Birkenau, Poland). In: Alvarado Sizzo, I. and López, A. (eds) *Turismo, Patrimonio y Representaciones Espaciales*. PASOS Revista de Turismo y Patrimonio Cultural, Colección Pasos Edita no.22, Tenerife, Spain, pp. 257–287.

Chávez, T. (2019) El turismo como instrumento de colonización neoliberal: una perspectiva foucaultiana (Tourism as an instrument of neoliberal colonisation: a foucauldian perspective). *Dimensiones Turísticas* 3(4), 110–116. DOI: 10.47557/MIAG2068.

Conanp (2020) Proclamación de la declaración Universal de los derechos de los animales (Universal declaration on the rights of animals proclamation). Available at: https://www.gob.mx/conanp/articulos/proclamacion-de-la-declaracion-universal-de-los-derechos-de-los-animales-223028 (accessed 12 January 2023).

Cooney, B. (2003) Posthumanity. In: *Thinking Philosophically About the Future*. Rowman and Littefield, Lanham, Maryland.

Crespi, M. and Domínguez, M. (2016) Los mercados de abastos y las crueldades turísticas (Supply markets and tourist cruelties). *PASOS, Revista de Turismo y Patrimonio Cultural* 14(2), 401–416. DOI: 10.25145/j.pasos.2016.14.026.

Encyclopedia of Life (2023) Ajolote, *Ambystoma mexicanum* (Shaw and Nodder 1798). Available at: https://eol.org/pages/1019571 (accessed 12 January 2023).

Espinoza, R., Chávez, R. and Andrade, E. (2012) Lo global-local como base del turismo y la sustentabilidad. Una aproximación a la discusión teórica conceptual (The global-local as a basis for tourism and sustainability. An introduction to a theoretical-conceptual discussion). In: López, A. (coord.). *Lo Glocal y El Turismo. Nuevos Paradigmas de Interpretación*. Universidad de Guadalajara/Academia Mexicana de Investigación Turística A. C., Guadalajara, Mexico, pp. 41–53.

Everett, S. (2008) Beyond the visual gaze? The pursuit of an embodied experience through food tourism. *Tourist Studies* 8, 337–358. DOI: 10.1177/1468797608100594.

Fennell, D. and Sheppard, V. (2020) Tourism, animals and the scales of justice. *Journal of Sustainable Tourism* 29(2–3), 1–22. DOI: 10.1080/09669582.2020.1768263.

Fernández, B. (2014) Antiguos tianquiztli, nuevos tianguis: cambios en los mercados y el comercio en la ciudad de México en el siglo XVI (Old tianquiztli, new tianguis: changes in the markets and trade in Mexico City in the sixteenth century). PhD dissertation, Universidad Complutense de Madrid, Madrid.

Foroudi, P., Cuomo, M.T., Foroudi, M.M., Katsikeas, C.S. and Gupta, S. (2019) Linking identity and heritage with image and a reputation for competition. *Journal of Business Research* 113, 317–325. DOI: 10.1016/j.jbusres.2019.09.042.

Francesca, F. (2019) *Philosophical Posthumanism*. Bloomsbury Publishing Plc, London.

Francione, G. (2010) Animal welfare and the moral value of nonhuman animals. *Law, Culture and the Humanities* 6(1), 24–36. DOI: 10.1177/1743872109348989.

Frew, E. and White, L. (2013) Exploring dark tourism and place identity. In: White, L. and Frew, E. (eds) *Dark Tourism and Place Identity. Managing and Interpreting Dark Places*. Routledge, Abingdon, UK, pp. 1–10.

García, B. (2017) Los mercados de abastos y su comercialización como producto del turismo de experiencias. El caso de Madrid (Supply markets and their commodification as tourism-of-experiences products. The case of Madrid). *Cuadernos de Turismo* 39, 167–189. DOI: http://dx.doi.org/10.6018/turismo.39.290491.

Gibson, C. (2012) Geographies of tourism. Space, ethics and encounter. In: Wilson, J. (ed.) *The Routledge Handbook of Tourism Geographies*. Routledge, New York, pp. 55–60.

Godoy, O. and Rodríguez, S. (2020) El no lugar y el antilugar a partir del imaginario del miedo en Cosalá, Sinaloa (the no-place and the anti-place from the imaginary of fear in Cosala, Sinaloa). *Topofilia, Revista de Arquitectura, Urbanismo y Territorios* 13(21), 308–335.

Gómez, G., Reyes, S., Teutli, C. and Valdez, R. (2007). La medicina tradicional prehispánica, vertebrados terrestres y productos medicinales de tres mercados del valle de México. (Pre-Hispanic traditional medicine, terrestrial vertebrates and medicinal products from three markets in the Valley of Mexico). *Etnobiología* 5, 86–98.

González, L. and Castañeda, R. (2012) ¿Geografía turística o geografía del turismo.? In: López, A., López, G., Andrade, E., Chávez, R. and Espinoza, R. (coord.) *Lo Glocal y El Turismo*, 1st edn. Academia Mexicana de Investigación Turística, A.C, Mexico City, pp. 54–73.

Hiernaux, D. and Lindón, A. (2006) Geografías de la vida cotidiana (Everyday life geographies). In: *Tratado de Geografía Humana*. Editorial Anthropos, UAM, Mexico City, pp. 356–399.

Hollinshead, K. (1999) Surveillance of the worlds of tourism: foucault and the eye-of-power. *Tourism Management* 20(1), 7–23. DOI: 10.1016/S0261-5177(98)00090-9.

Lennon, J. and Foley, M. (2000) *Dark Tourism. The Attraction of Death and Disaster*. Thomson, London.

Linares, E. and Bye, R. (2013) Códice de la Cruz-Badiano. *Arqueología Mexicana,* special edition 50.

López, A. and Van Broeck, A.M. (2013) Metodología para el abordaje del trabajo sexual masculino en localidades turísticas de México (A methodology for the study of male sex-work in tourist locations in Mexico). In: López, A. and Van Broeck, A.M. (eds) *Turismo y Sexo En México. Cuerpos Masculinos En Venta y Experiencias Homoeróticas*. Geography Institute, National University of Mexico, Mexico City, pp. 107–132.

López, A. and Quintero, G. (2021a) Animal dark tourism in Mexico: bulls performing their own slaughter. In: Rickly, J. and Kline, C. (eds) *Exploring Non-Human Work in Tourism*, Vol. 5. pp. 69–82. DOI: 10.1515/9783110664058.

López, A. and Quintero, G. (2021b) La geografía del turismo y la geografía de los animales intersectadas por la ética poshumanista (Tourism geography and animal geography intersected by posthumanist ethics). *Cuadernos de Geografía: Revista Colombiana de Geografía* 30(1), 86–105. DOI: 10.15446/cdg.

Low, P. (2012) The Cambridge declaration on consciousness. In: *Francis Crick memorial conference on consciousness in human and non-human animals*, Churchill College, Cambridge University, Cambridge, UK, July 7, 2012. Available at: https://fcmconference.org/img/CambridgeDeclarationOnConsciousness.pdf (accessed 28 April 2021).

Macdonald, S. (2009) *Difficult Heritage. Negotiating the Nazi Past in Nuremberg and Beyond*. Routledge, Abingdon, UK.

Marres, N. and Weltevrede, E. (2013) Scraping the social? *Journal of Cultural Economy* 6(3), 313–335. DOI: 10.1080/17530350.2013.772070.

Martínez, A. (2010) Venta de animales en el mercado de Sonora. Una radiografía reportaje (The sale of animals in the Sonora market A radiography-report). Undergraduate Thesis, National University of Mexico, Mexico City.

Mata, A. (2020) Hacinamiento en mercado de Sonora por contingencia (Cramming at the Sonora market under confinement). *Excelsior*. Available at: https://www.excelsior.com.mx/comunidad/hacinamiento-en-mercado-de-sonora-por-contingencia/1374754 (accessed 28 April 2021).

Méndez, E. (2012) De anti-lugares, o la difusión de la narcoarquitectura en Culiacán (Of anti-places, or the diffusion of Narco-architecture in Culiacan). *URBS. Revista de Estudios Urbanos y Ciencias Sociales* 2(2), 43–62.

Mexico City's Government Gazette (2020) *Gaceta Oficial de la Ciudad de México* 21, 402.

Mkono, M. and Holder, A. (2019) The future of animals in tourism recreation: social media as spaces of collective moral reflexivity. *Tourism Management Perspectives* 29, 1–8. DOI: 10.1016/j.tmp.2018.10.002.

Neumann, J.M. (2012) The universal declaration of animal rights or the creation of a new equilibrium between species. *Animal Law* 19, 63–109.

Ortiz Millán, G. (2020) Pandemias, zoonosis y comercio de animales silvestres (Pandemics, zoonosis and the trade of wild animals). *Perspectivas Bioéticas. Revista de Bioética y Derecho* 50, 19–35. DOI: 10.1344/rbd2020.50.31303.

Portilla, A. (2002) Una estrecha relación entre el turismo, la geografía y el mercadeo (A close relationship between tourism geography and trade). *Geoenseñanza* 7(1–2), 109–113.

Prats, L. (1998) El concepto de patrimonio cultural (The concept of cultural heritage). *Política y Sociedad* 27, 63–76.

Quintero, G. and López, A. (2017) Tauromaquia y turismo oscuro en México: las corridas de toros como prácticas no éticas (Bullfighting and dark tourism in Mexico: bullfights as unethical practices). *Teoría y Praxis* 24, 197–228.

Quintero, G. and Castro, A.M. (2022) La creación del branding turístico y de la sinécdoque turística de Zacatlán a partir de la descampesinización local y el comercio de los productos de manzana (The creation of tourism branding and the Zacatlan tourism synecdoche from local depeasantisation and the trade of apple products). *Revista Ayana* 2(2), 1–17.

Rico, T. (2008) Negative heritage: the place of conflict in world heritage. *Conservation and Management of Archaeological Sites* 10(4), 344–352. DOI: 10.1179/135050308X12513845914507.

Rodríguez, C. (2022) *Hipermétodos: Repertorios de La Investigación Social En Entornos Digitales (Hyper-Methods: Repertoire of Social Research in Digital Environments)*. Metropolitan University of Mexico (UAM), Mexico City. DOI: 10.24275/9786072824812.

Sarukhán, J. (2009) ¿Por qué se pierde la biodiversidad? (Why is biodiversity lost?). Interview on video. Conabio. Available at: https://www.biodiversidad.gob.mx/biodiversidad/porque (accessed 12 January 2023).

Semarnat (2017) General law of wildlife, conservation and sustainable use – Ley general de la vida silvestre, conservación y aprovechamiento sustentable. Gobierno de México. Available at: https://www.gob.mx/semarnat/articulos/ley-general-de-vida-silvestre-conservacion-y-aprovechamiento-sustentable (accessed 12 January 2023).

Sharpley, R. (2022) Tourist experiences. Liminal, liminoid or just doing something different? In: Sharpley, R. (ed.) *The Routledge Handbook of the Tourist Experience*. Routledge, Abingdon, UK, pp. 89–100. DOI: 10.4324/9781003219866.

Singer, P. (1975) *Liberación Animal*. Taurus, Spain.

Smith, J. (2003) Beyond dominance and affection: living with rabbits in post-humanist households. *Society & Animals* 11(2), 181–197. DOI: 10.1163/156853003769233379.

Tena, R. (2020) Los centros históricos y la compleja relación patrimonio-turismo (Historic centres and the complex heritage-tourism relationship). In: González, C. and Hiernaux, D. (eds) *Patrimonio, Economía Local y Turismo En Centros Históricos*. Universidad Nacional Autónoma de México, Mexico City, pp. 17–33.

TripAdvisor (2022) Mercado de Sonora. Available at: https://www.tripadvisor.com.mx/Attraction_Review-g150800-d152194-Reviews-Mercado_Sonora-Mexico_City_Central_Mexico_and_Gulf_Coast.html (accessed 12 January 2023).

Tuan, Y. (1974) *Topophilia: A Study of Environmental Perception, Attitudes, and Values*. Columbia University Press, New York.

Tuan, Y. (1976) Humanistic geography. *Annals of the Association of American Geographers* 66(2), 266–276. DOI: 10.1111/j.1467-8306.1976.tb01089.x.

Tuan, Y. (1994) *Espacio y Lugar. La Perspectiva de la Experiencia*. Universidad Nacional del Sur, Bahía Blanca, p. 1. Trad. Yezzi.

Tunbridge, J.E. and Ashworth, G.J. (1996) *Dissonant Heritage. The Management of the Past as a Resource in Conflict*. Wiley, Hoboken, New Jersey.

UNWTO (2023) Código Ético Mundial para el Turismo. Available at: https://www.unwto.org/es/codigo-etico-mundial-para-el-turismo (accessed 12 January 2023).

Urry, J. and Larsen, J. (2011) *The Tourist Gaze 3.0*. New Delhi. DOI: 10.4135/9781446251904.

6 *Jallikattu*: Walking the Tightrope between Culture and Animal Welfare

Javed Salim* and Nusrat Yasmeen
Jamia Millia Islamia (A Central University), New Delhi, India

Abstract
Cultural traditions constitute a substantial element of the tourism offer. This chapter examines the usage of non-human animals to commemorate a sacred tradition, namely bullfights in India. For this purpose, a case study of *jallikattu* has been undertaken. The chapter highlights the engagement of multiple actors, including the community, the judiciary, NGOs, animal rights activists and environmentalists, and the ongoing conflict between citizens and the state due to their divergent viewpoints. Consequently, economically developed nations are likewise attempting to reduce the repercussions on non-human animals by abolishing these cruel rites. The ongoing conflict between cultural rights and animal rights renders the posthumanist goal of erasing the human–animal divide nearly impossible.

Introduction

The interaction between humans and animals has developed since time immemorial, from humans as hunters and animals as their prey, or vice versa. The domestication of animals by humans in agriculture, hunting and transportation has gone through different ages (Farooqui, 2001) and the relationship between human and non-human animals has been changing continuously, giving way to new forms of engagement. Non-human animals have been commodified and commercialized in various industries and have been introduced into many cultural practices with both positive and negative consequences for them (Markwell, 2015).

This chapter focuses on how non-human animals are involved in heritage and cultural tourism. For ages, the practice of using animals for private and public entertainment has gone unchallenged and been deemed acceptable. The modern tourism business has increased and enabled animal availability and access, allowing tourists to interact with a diverse range of species all over the world. Recently, along with many other industries that use animals, the ethics of animal use in entertainment/tourism is being questioned. Animal ethics is concerned with the moral status of individual animals and the consequences of their actions; an ethical stance toward animals would imply that humans should consider their sentience and interests (Winter, 2020).

Jaume Guia (2020) argues that posthumanism is the only ethical theory that can fight global commodification, restore politicization (as again apolitical moralism) and bring fundamental change and a real alternative to globalization, because it anticipates subjectivities or identities

*Corresponding author: javed.jmi007@gmail.com

 Tourism, Heritage and Commodification of Non-human Animals (eds A. López López *et al.*)
DOI: 10.1079/9781800623309.0006

other than those of humanism. Posthumanism strives to eliminate the human–animal divide by denying human exceptionalism – founded on humanist philosophy – and destabilizing the core ontological concepts that have dominated Western thinking from the Enlightenment to the present. While modernist tourist studies prioritize the visitor's behaviour, attitudes and views, postmodernist studies argue that tourism should be researched holistically, including non-human species' interests and perspectives (Cohen, 2019).

The role of animals in non-consumptive wildlife tourism is significant, as highlighted by Moorhouse *et al.* (2016), with estimates suggesting that they contribute to between 20 and 40% of all international travel. However, animal welfare has traditionally not received much attention within this industry. In recent years, scholars have started to raise concerns about non-human animal rights in tourism. Caroline Winter conducted a thorough analysis of around 540 articles related to animals in tourism and found that although animals play a crucial role in this sector, they are often not accorded the respect and recognition they deserve (Winter, 2020). Sheppard and Fennell conducted a content analysis of tourism policies in 73 countries, which showed evidence of changes through time and concluded that 'the policies reveal an instrumental level of concern for animal welfare' (2019, p. 139) and are centred on providing benefits for humans only.

Our interest here is on the degree to which animals are abused for the sake of heritage and culture. We underline the stances of various organizations and governments on animal abuse and the collaboration of different actors (activists, lawyers, environmentalists, celebrities and others). Some are affiliates of international groups, such as PETA (People for the Ethical Treatment of Animals) and WWF (World Wildlife Fund), or national organizations, such as the AWBI (Animal Welfare Board of India, a statutory body set up by an act of Parliament) and FIAPO (Federation of Indian Animal Protection Organisation). These organizations (many of them non-profit) are dedicated to freeing, treating and nourishing animals, and some of them have been pursuing lawsuits to get the court to stop animal abuse and other inhumane activities and legalize animal rights (Berti, 2019).

The 'dilemma' posed by animal protection in the West, where animals are simultaneously perceived as resources for human consumption and 'subjects of empathy', is found in post-colonial India to be a more complicated scenario, given the country's political and socio-religious dynamics (Chakrabarti, 2010). Animal cruelty, and particularly animal sacrifices, have been questioned in India by various philosophical and religious movements that promote ideas such as non-violence, reincarnation and purity, to name a few (Berti, 2019).

This chapter presents a case study from India. It shows how a popular sport involving animal suffering (*jallikattu* or bullfighting) attracted the attention and active participation of different actors, including external ones, to stop the pain and suffering of bulls and put an end to an illicit affair that has been in practice for decades.

A bullfight, in which the bull is usually killed, has profound origins in Mediterranean history. Yet, any comparison of modern-day demonstrations to ancient ones must take into consideration the vast social variations between the periods (Shapland, 2013). In every region of the world where it is still practised, bullfighting is celebrated differently. However, it is usually considered a sacred ritual having heritage and cultural value (Maudet, 2006). Erik Cohen specified the two major forms of bullfighting, i.e. animal vs animal contests, in which animals fight one another in human-designed and managed environments, and human vs animal contests, in which people fight animals, the latter form generally being considered more lethal than the former one (Cohen, 2014).

Bullfighting is permitted in eight European and Latin American countries (France, Portugal, Spain, Mexico, Venezuela, Peru, Ecuador and Colombia) and in the USA and India, with some adaptations (Hoyt, 2017). This study seeks to learn from other examples and suggests new measures to ensure the well-being of non-human animals in such practices.

Development of Research

Sheppard and Fennell (2019) studied 123 tourist policies from 73 nations to assess the state of the use of animals in them. Outstanding among these documents are Tanzania's 1999 National Tourism Policy, which clearly defines economic, social, cultural and environmental goals. This policy appears to classify wildlife mostly as a tourist attraction. South Africa's 2009 City of Cape Town's Responsible Tourism Policy, which states that 'No captive wildlife is held, except by those authorised and suitably equipped to house and care for them' (City of Cape Town, 2009, p. 34), and Ghana's 2013–2017 National Tourism Development Plan, mainly focused on protecting wildlife and marine life in designated sanctuaries. In contrast, other nations seem to encourage consumptive and perhaps exploitative animal use. For example, *Gaining the Edge 2015–2018*, the tourism plan for British Columbia, promotes hunting and fishing as tourist activities, *Towards Sustainable Travel and Tourism in Norway: A road map* portrays a person holding a dead grouse by its feet (labelled as 'local cuisine'), a reindeer hauling a sleigh of, presumably, visitors and a fisherman holding a fish with its mouth open to expose the enticement, and many other examples. With the *Declaration of Toulon*, which recognizes animals as persons who should be treated equally, France has advanced in its efforts to promote animal care (Regad and Riot, 2019). However, despite apparent progress in tourism policy over the years, none of the reviewed policies stands out as a model for identifying and/or addressing the negative consequences of tourism activities on animal welfare. To expect that tourist policy would ultimately include human and animal well-being and rights is outright optimistic and even naive (Sheppard and Fennell, 2019).

The concepts of animal rights and animal welfare are very different from each other. The cage analogy can be used to visualize their differences: animal welfare entails larger cages and animal rights entail empty cages. The majority of animal ethics studies in the area of tourism concentrate on animal welfare rather than animal rights (see Fennell, 2012, 2013).

A tourism policy is influenced by variables such as social, cultural and ecological circumstances (Dredge and Jamal, 2015). Therefore, various stakeholders, such as developers, resource managers and communities have their individual objectives and want to defend their interests, and hence the existing gap between policy approval and policy execution.

Animals pay a high price for amusing the audience and generating profit for the industry: they are hunted and trapped in the wild, caged indefinitely, used as transportation, consumed, brutally trained, molested, killed on purpose (Hughes, 2001; Carmeli, 2002; Shani and Pizam, 2008; Cohen, 2013, 2014) and even raped (Reuters, 2015). Animals are confined to cages, drugged, restrained by way of ropes, controlled through brutal procedures and forced to adopt strange behaviour, all of which denies their inborn needs and injures them. Many animals exploited in the tourism business are in precarious conditions from which they will never be able to escape, and there are numerous incidents of their mistreatment and cruelty (Winter, 2020). Animals are employed for amusement and commerce in all of these scenarios (on the commodification of animals, see Wearing and Jobberns, 2010). However, a group of tourism scholars, based on ethical principles that can help answer questions about whether or not such use is justifiable, is raising concerns about the use of animals in tourism (Winter, 2020).

According to Tom Regan (2004), individual animals have intrinsic value, just like humans. The fact that animals are mindful, deliberate in their acts and ideas, and sentient (which implies they feel emotions like pain and pleasure) is one of the criteria to support this position. Regan's focus is deontological and imitative of democratic formal justice, according to which no one is better than the other and hence does not deserve better treatment. It is predicated on respect and the same level of regard (Regan, 2004). Non-human species have different needs, which should be satisfied by tour operators, tourists and all other tourism shareholders (Fennell, 2013). However, animal treatment varies within different civilizations. Certain animals that are regarded holy or special in some faiths and cultural groupings are notable examples. Therefore, animal welfare, and what is considered suitable or unsuitable in terms of human and non-human animal relationships, differs greatly among cultures.

The idea of cultural relativism, which holds that some acts should be free from criticism, could lead to a confrontation between tourists and local traditions, as well as between ethical perspectives. There are various conflicting cultural associations between human and non-human animals that have existed for decades and are still practised, such as bullfighting, whaling, cockfighting, dogfighting, etc. In this chapter, the focus is on bullfighting. Bullfighting, according to Cohen, is particularly problematic, as it pits animal ethics against long-standing cultural practices that generate significant local revenue (Cohen, 2018).

An Indian Context

According to pro-animal welfare thinkers in India, non-human animals always had a distinct position in Indian history. They refer to Indian traditional kindness towards these animals, reflected in the idea of *ahimsa* or non-violence and the 'emblem of *Asoka*', which is in true spirit compassion, kindness and justness to all. The idea of compassion is fundamental to animal welfare establishments in India. For example, 'Compassion in action' is the slogan of the AWBI, an institution set up by Pandit Jawaharlal Nehru in 1962 (Berti, 2019). Chinny Krishna, an animal activist and vice-chairman of AWBI argued how India as a nation 'carries with her a five-thousand-year-old holy inheritance of a rich, spiritual tradition of *ahimsa*, or non-injury to other sensitive beings, and has always recognized the existence of all living beings' (Animal Welfare Board of India, 2012, p. 5). Krishna also used the concept of *karma* and the possibility of animals reincarnating as humans to defend what he calls an 'inherent aspect of our country's religious beliefs' that 'causing violence to animals has always been disdained as an act of immense sin' (Animal Welfare Board of India, 2012, p. 5).

The animal welfare and rights school of thought would come later to the Indian subcontinent through indefinite channels. During the colonial era, British rulers in India would play a confusing role. On the one hand, they produced legislation to combat animal cruelty (the first law being passed in Bengal in 1869), while on the other they substantially abused animals in colonial economic ventures such as the army and on hunting excursions. Some British colonists began to feel compassion for native animals and the first Indian societies for the prevention of cruelty to animals were formed (Chakrabarti, 2010). At the same time, some religious reform organizations, like the Arya Samaj, started participating in the animal welfare movement debate for cow protection and denoted the cow as a symbol of unification for Hindus. In his works, Swami Vivekananda, founder of the Ramakrishna Mission, used to state that 'from the tiniest worm crawling beneath our feet to the most powerful beings who have ever existed, all have different bodies but one soul' (quoted in Gosling, 2001, p. 39). Others made a theological argument, claiming that 'it is our obligation to God, who created animals, to treat them with kindness' (Lord Erskine, quoted in Blosh, 2012, p. 30).

During the later phase of the colonial period – the early 1900s – a new fashion started to emerge, namely, the transition of hunters into conservationists for flora and fauna as an after-effect of the development of animal rights movements in the West. Hunting would lose some of its lustre, and the protection of (at least certain) animals became the new symbol of society (Rangarajan, 1998; Rajamannar, 2012). Jim Corbett, an Indian-born British hunter who is generally portrayed as a revolutionary player in the movement that transformed hunters (both colonists and Indians) into conservationists-turned-photographers, is a name that comes up frequently. Concerns against animal mistreatment prompted the shift, as well as the 'increasing recognition that unless rigorous approaches were adopted, entire species may vanish from India' (Rajamannar, 2012, p. 125). Recently, the Himachal Pradesh government renamed the Simbalwara National Park in Sirmaur as Sher Jung National Park. Sher Jung was also a well-known hunter-turned-conservationist who, like Corbett, wrote several books about the forest, animal behaviour and hunting (Berti, 2019). Above all this, then Prime Minister of India, Indira Gandhi, had a passion for wildlife which can be seen in her policies for wildlife security. Not only is her name associated with the world's first large-scale wildlife conservation programme, 'Project Tiger', but also it was under her regime that the Wildlife (Protection) Act was approved in 1972, which efficiently debarred hunting in India and paved the way for the establishment of protected areas such as national parks and wildlife sanctuaries (Ramesh, 2018).

The 'animal cause' has grown in popularity, to the point that now it is a prominent topic. PETA recently enlisted Bollywood celebrities to model for a bold and colourful commercial campaign, with each advert giving out a distinct message: 'Help the birds by being an angel. Don't cage them', says actor John Abraham, who stands with huge, mighty outstretched wings; 'Let vegetarianism grow on you', says former Miss Universe and actress Lara Dutta, clad in body-hugging attire made of vegetable leaves; and, in opposition to the skin trade, actress Dia Mirza poses in bloody leopard skin (Berti, 2019).

Jallikattu: A Case Study

Bullfighting, or 'human-initiated agonistic animal competitions' (Cohen, 2014) with or between animals have been introduced into tourism studies as the prototypical, most popular, yet controversial example of a widely diverse class of human and non-human animal relationships. These competitions come in two major subclasses, as mentioned earlier. Bullfighting, as we know it today, is based on traditions that developed in the Iberian Peninsula over the last three or four centuries (Cohen, 2014). Many countries, including Canada, Cuba, Argentina, Italy, Denmark and the UK, have already made bullfighting illegal, but it is still practised in some nations around the world (Humane Society International, 2020), and with some variations in the USA, India, Turkey, Bosnia and the Persian Gulf.

The case of *jallikattu*, the South Indian version of bullfighting, provides an example of how far people can go to defend practices they consider part of their culture. Human and non-human animal engagement in India's culture is not new and is not confined to bullfights. Cockfights, *kambala* or buffalo races, *bailgadi shariat* or bullock cart races, camel races, dog fights, *bulbul* (Indian nightingale fights) and many other forms continue to exist despite several attempts from the authorities to put a halt to animal exploitation. These events are observed annually with or without consent from the authorities (Kumar, 2017; Kutubuddin, 2017).

A few years ago, *jallikattu* attracted worldwide attention due to the agitation its attempted prohibition provoked in the state of Tamil Nadu. In what follows, we shall look at what exactly *jallikattu* is, the involvement of activists and organizations from all over the world, its historical and cultural significance that led to the mass movement against the highest-level authorities, and the current position this sport occupies now in India.

Also known as *eru tazhuvudhal* or *manju virattu*, *jallikattu* is a traditional annual rural sport conducted during Pongal, the four-day-long harvest festival celebrated in mid-January in Tamil Nadu, South India. It is a festivity dedicated to offering thanks to nature. The third day, known as Mattu Pongal, is dedicated to cattle on account of the important role they play in Indian agriculture, and it is on this day that *jallikattu* takes place. It is an attempt to hold on to the bull's horns or hump for a given time within an enclosed arena. *Jallikattu* occurs in various ways. Historically, it aimed to demonstrate men's mastery of the bull by attempting to remove a cloth tied to one of its horns. This could be done by a single individual or in small groups (Hart, 2014).

Jallikattu is a compound term formed by the words *jalli/salli* ('coins') and *kattu* ('tie'), since the cloth thus tied contained coins that the men gathered in the arena attempt to grab. In that sense, it is not so much about 'bull-taming' as it is about the 'male bravery' involved in seizing a prize from a bull's head that's been let loose in the arena (Kalaiyarasan, 2017). This adrenaline surge for displaying masculinity is similar to the one experienced by the *torero* during a *corrida* in Spain (López-López and Quintero Venegas, 2021). According to mythology, ladies preferred men who were victorious in *jallikattu* as their spouses. Today, *jallikattu* is sponsored by powerful men in the villages, but in the past it was patronized by a monarch (Brara, 2019).

Jallikattu dates back as far as the Indus Valley civilization 5000 years ago (Joshi and Parpola, 1987). The brutal sport is described in detail in the *Sangam* literature of Tamil Nadu, which contains the most essential information about Tamil history and culture (Sen, 1999). *Jallikattu* is also mentioned in the comprehensive papers of British officials, who emphasize the cultural significance of the sport (Thurston and Rangachari, 1909), as well as in the writings of renowned historian A. L. Basham, who claims that it has 'some sacred importance' in Dravidian culture (Basham, 2004).

The bulls used are indigenous Indian cattle breeds, whose preservation and long-term viability are inextricably linked to the sport's survival. Kangayam, Pulikulam, Umbalachery, Barugur and Malai Maadu are some of the native breeds employed in the *jallikattu* sport. These bulls are robust, active, have a well-developed hump and are more immune to communicable and parasitic diseases than crossbreeds in the area (Vivekanandan, 2013).

According to the Ireland-based NGO Animal Rights Action Network (ARAN), *jallikattu* is a violent contact sport in which terrified bulls are encircled by hundreds of screaming men, get punched in the face, have their tails twisted and tugged – and in some cases broken or torn off – and are pounced on and knocked to the ground. Men, sometimes alone, sometimes in groups, leap on the backs of bulls and attempt to drag them to the ground. Chilli powder is smeared into the bulls' eyes, they are given alcohol and then chased by crowds who try to overwhelm and bring them down (Suroor, 2012).

Pursuing Lawsuits

In 2006, the son of a local Nagaraja died while sketching the sporting event of *jallikattu*. The Nagaraja filed a petition in the Madras High Court against what was called an ill-form of animal sport. As a result, the court barred all types of bull-related sports or entertainment (Imranullah, 2017). Later, however, the Madurai Bench of the High Court decided to allow the sport on condition that the safety of animals, contestants and onlookers be ensured. In 2009, the state government enacted the Tamil Nadu Regulation of Jallikattu Act, giving the green light to the practice of the sport with certain precautions and regulations. But 2 years later, the central government in New Delhi issued a notification to ban the use of bulls in sports performances (Jayashree *et al.*, 2019). Nevertheless, and after all these legal initiatives, *jallikattu* remained in practice. It was then that PETA entered the scene to denounce the infringement of the law and strongly encourage pro-animal welfare movements. Things got critical when the AWBI produced photographic and video evidence showing the ill practices of entertainment of this animal sport to the Honourable Supreme Court of India, the highest judiciary in the country, which in 2014 completely banned the conduct of bull-related sports, including *jallikattu* in Tamil Nadu, bullock-cart races in Maharashtra and other such events, citing them as violations of several sections of the Prevention of Cruelty to Animals Act (PCA) of 1960 (Sharma and Singh, 2015). A dispute ensued, in which PETA was identified as the principal opponent of 'Tamil culture', a 'foreign influence' that was responsible for the ban (Jayashree *et al.*, 2019).

To Tamil people, this ban on *jallikattu* was an attack on their culture and pride. The sense of their culture being under threat cut through the rural–urban divide, uniting these disparate groups under the banner of bullfighting. *Jallikattu* became associated with Tamil pride, and as popular opposition to the ban mounted, supporters of the ban were labelled as anti-Tamil. The topic served as a platform for constructing and defining Tamil identity (Sarukkai, 2017). Other contemporary political and social issues were combined with the ban to increase public outrage. Farmers' economic distress, as well as an increase in farmer suicides owing to crop failure as a result of one of the worst droughts in Tamil Nadu in decades, fuelled populist emotions of Tamil identity (Lal, 2017).

A revolution erupted across the state of Tamil Nadu, which was regarded as extraordinary in its intensity and widespread support. Pro-*jallikattu* protests to reintroduce the sport multiplied everywhere. The protests were described as 'iconic' in Tamil Nadu history for not only were they big, non-violent and spontaneous, but they also drew people from all walks of life, with a large number of women and men, largely young people. Surprisingly, the cities, particularly Chennai, the state capital, were the main locus of protest, an unusual occurrence because *jallikattu* is typically associated with agricultural landscapes and harvest festivals in rural areas. *Jallikattu* is not held in cities, and the city people seen at the event are tourists who are witnessing it (Jayashree *et al.*, 2019). On 20 January 2017, an estimated 100,000 demonstrators congregated on Marina Beach alone (*Deccan Chronicle*, 2017), and public transportation was overflowing with people heading to the protest sites. According to certain sources, the number of individuals ranged from half a million to a million (Moorthy, 2018). Protests included more than simply rallies and chants. Students enacted *jallikattu* by stretching out their hands atop their heads to represent the bull's horns, getting themselves painted as bulls and performing folk dances while traditional instruments such as drums were played for youth who danced to the beats for a two-kilometre stretch, according to news reports (*The Economic Times*, 2017).

Another controversy took place between the two contradictory views on *jallikattu* by the Union government and by the AWBI. While the former wanted to let people celebrate this cultural festival – probably anticipating a few electoral gains in the poll-bound state of Tamil Nadu – the latter desired to put an end to this cruel practice on non-human animals. The Union government attempted to bypass the Supreme Court's 2014 verdict by issuing a statement authorising the sport, against which AWBI filed writ petitions in several courts in pursuit of directions or relief against the Government of India when it had explicitly been instructed to refrain from doing so (Ravishankar, 2016). R. M. Kharb, chairman of AWBI, openly criticized the political choices based on electoral gain that proved damaging to the well-being of animals, who have 'no voice and no vote' (Vishnoi, 2016). After this, Mr Kharb and his associates were asked to submit their resignation to the Government (Ravishankar, 2016).

Parallel to this, a review petition was filed in May 2014, which was dismissed in 2016. After witnessing so much ferocity and the agony of Tamil people for the conservation of their culture and pride, in 2017 the Tamil Nadu government adopted the Prevention of Cruelty to Animals (Tamil Nadu Amendment) Act and the Tamil Nadu Prevention of Cruelty to Animals (Conduct of Jallikattu) Rules to overturn the Supreme Court's 2014 verdict. The Amendment Act of 2017 defined and legalized *jallikattu*, based on its being part of culture, and encouraged the protection of indigenous cattle breeds. *Jallikattu* rules were established for taking care of the bulls involved (Thakur, 2018). As the case is still pending in court, the conflict between both parties continues.

What was by tradition filled with fun and gaiety has now been transformed into a big-money sport. Bulls are especially skilled and are poked to run when they baulk at the sight of a cheering crowd (Kumar, 2017). Foreign tourists flock to see bullfights, drawn by its image in popular literature and culture, even though *jallikattu* was never really a chief attraction for the vast majority of tourists visiting Tamil Nadu, despite the promotion by state tourism authorities of a *Jallikattu* Tour (Tamil Nadu Tourism, n.d.). The growing disagreement with bullfighting by animal rights activists and its growing criticism as a barbaric or uncivilized custom has affected foreign tourist turnout at fights. One of the prominent examples is that of Irish groups declaring to launch a Europe-wide campaign for a 'tourism boycott' of India to dissent against *jallikattu*, expressing it as a 'cruel bull-abusing contact game' (Suroor, 2012).

The conflict between animal rights and cultural rights features in laws all over the world. Mitigating measures drawn from other experiences may prove helpful to the Indian bullfighting scenario.

In Spain, bullfighting plays an important – albeit contentious – role in tourism. The bull is a national symbol of the country as well as a landmark in its countryside (Brandes, 2009). Bullfighting is a large business that has grown significantly as a result of the tourism industry's demand. It was initially conducted mainly at carnivals, but from the 18th century onwards, as it was adopted by the rising urban classes, bullfighting came to be practised outside traditional festivities as an autonomous, commercialized activity, carried out by various professional *toreros*, and oriented towards a massive public made up of people from all social classes (Cohen, 2013). While bullfighting was outlawed in the north-eastern Autonomous Community of Catalonia in 2010, in 2013 two statutes at the national level classified it as 'cultural heritage', and the restriction in Catalonia was repealed by the Spanish Constitutional Court in 2016. It was noted that the State's authority to preserve 'cultural heritage' takes priority over the autonomous community's statute prohibiting the activity. The Court, on the other hand, agreed that the autonomous community could regulate the organising and development of bullfighting events (Burgen, 2016).

Likewise, in France, any act of cruelty towards an animal is punishable under the Criminal Code. Traditional bullfighting regions, however, are exempt from the rule. The exemption for a ban on bullfighting was challenged by animal rights advocates, but the French Constitutional Council dismissed the petition (Constitutional Council, 2012). This exception was deemed reasonable by the French Court. Even though bullfighting is recognized as a cultural heritage and a traditional practice in both Spain and France, the practice has not been regulated under the rationale of cultural rights. This is because cultural rights have not received the same level of protection as human fundamental rights in these countries (Thakur, 2018).

In South Africa, cultural rights are protected by Article 31 of the Constitution. During the first Fruit Festival, a case was presented before the High Court against a traditional ritual in which young soldiers slaughter a bull with their bare hands. According to the High Court, what was at stake were the constitutional rights of the Zulu people to pursue their religion and culture. The application to outlaw the practice was denied because it prioritized cultural rights over animal protection. The importance of tradition in the lives of Zulu people was emphasized, leading to the conclusion that the balance swings heavily in favour of the traditional community (Smith, 2012). In the UK, however, a different stand is taken. Though fox hunting as a cultural practice is profusely depicted in art and literature, the Hunting Act 2004 prohibited it, in an act that many regarded as a breach of the European Convention on Human Rights (ECHR). A claim was submitted and dismissed, and the activity remains illegal (UK Court, 2007).

Disputes between cultural and animal rights activists are addressed differently across nations. Cultural rights have been given a significant position in certain countries' constitutions and confirmed by the judiciary. As Cohen (2019) points out, posthumanism's extreme ethical inclusion is appealing, but its efforts to place humans and non-human creatures on the same level appear to be plagued with ambiguities. Even if non-sentient animals and germs are excluded, posthumanist discourse in broad terms continues to have trouble conceptualizing major distinctions across animal groups. Posthumanism may not be widely accepted in the near future. Further, in some nations, animal welfare legislation makes an exception for cultural and customary behaviours. Where the Constitution or animal welfare legislation does not provide a clear position, disagreements are addressed by the country's judiciary. Simultaneously, to resolve these disagreements, the courts of these countries have upheld that such behaviours can be regulated. As a result, the courts have ruled in favour of cultural activities that are restricted (Thakur, 2018).

Conclusion

On a review of the relevant literature, it appears that participants in the *jallikattu* sport have the intention of stopping the bull by embracing rather than dominating it. Deeply ingrained in Tamil culture, *jallikattu* holds immense cultural significance for the people of Tamil Nadu. Holistic development for all is an ideal that has been around since the birth of civilization. If this ideal is ever to be attained, a balance must be maintained between culture preservation and animal welfare. It is up to the State then to recognize and promote the cultural significance of animals and to ensure that they can lead a life free of the pain, tension, stress and agony they have endured over the years.

The farming of stud bulls has a complex relationship with traditional farming, as farmers favour the winners of *jallikattu* competitions. These bulls' calves are in high demand since they will be employed as draught animals in transportation and agriculture. *Jallikattu* is now caught in a fight between animal rights and the protection of Indian culture. This chapter illustrates an ongoing debate on the conflict between cultural rights and animal rights that has no end in sight. Since the advent of animal rights and welfare schools of thought, bullfighting has been a scrutinized event because of its close association with the cultural sentiments of a particular region or community. Many nations have attempted and failed to curtail such inhumane practices, as cultural and traditional beliefs have always stood in the way.

In the case of *jallikattu*, the anti-ban campaign claims that this 'country sport' represents guts, drive and willpower, and supports its allegation on photographic proof. The ban supporters doubt the bravery involved, claiming that there is little difference between taming and torture. While one faction portrays the event as a pleasant race, the other camp portrays it as a bloodbath intent on maiming the helpless animal. Whatever the case, the fact is that the animal has no say in the situation. *Jallikattu* supporters defend it as 'a native tradition' and look forward to its preservation. Without it, they claim cultural identity would wane.

Be that as it may, in India bullfighting has become a huge business that has grown substantially as a result of tourism demand. Despite the controversy, bullfights remain popular tourist attractions. The act of taming a bull by 'embracing' it – that is, grasping its tail or horns while clinging to its body – is a cruel deed in itself. Furthermore, who will be responsible for determining how long the taming lasts and how will this be enforced? If it goes beyond the limits, what measures are to be taken? Who would be held liable if there were any unfortunate situations involving people (including tourists) or bulls? How to create a balance between cultural beliefs and tourism products in the case of non-human animals? Or, is it even possible to create a balance?

References

Animal Welfare Board of India (2012) *Animal Citizen*. Animal Welfare Board of India, Chennai, India.

Basham, A. (2004) *The Wonder That Was India*. Picador, UK.

Berti, D. (2019) Animals in the public debate: welfare, rights, and conservationism in India. *Religions* 10(8), 475. DOI: 10.3390/rel10080475.

Blosh, M. (2012) The history of animal welfare law and the future of animal rights. Master's thesis, The University of Western Ontario, Ontario, Canada, pp. 1–127.

Brandes, S. (2009) Torophiles and torophobes: the politics of bulls and bullfights in contemporary Spain. *Anthropological Quarterly* 82(3), 779–794. DOI: 10.1353/anq.0.0076.

Brara, R. (2019) Animal rights vs. bullfights: the horns of an Indian dilemma. Available at: https://www.environmentandsociety.org/arcadia/animal-rights-vs-bullfights-horns-indian-dilemma (accessed 2 March 2022).

Burgen, S. (2016) Spanish court overturns Catalonia's bullfighting ban. Available at: https://www.theguardian.com/world/2016/oct/20/spanish-court-overturns-catalonia-bullfighting-ban (accessed 10 March 2022).

Carmeli, Y.S. (2002) 'Cruelty to animals' and nostalgic totality: performance of a travelling circus in Britain. *International Journal of Sociology and Social Policy* 22(11/12), 73–88. DOI: 10.1108/01443330210790201.

Chakrabarti, P. (2010) Beasts of burden: animals and laboratory research in Colonial India. *History of Science* 48(2), 125–152. DOI: 10.1177/007327531004800201.

City of Cape Town (2009) Responsible tourism policy for the city of Cape Town. The City of Cape Town Tourism Department. Available at: https://resource.capetown.gov.za/documentcentre/Documents/Bylaws%20and%20policies/Responsible%20Tourism%20Policy%20for%20the%20City%20of%20Cape%20Town%20-%20approved%20on%2026%20November%202009.pdf (accessed 26 July 2023).

Cohen, E. (2013) 'Buddhist compassion' and 'Animal abuse' in Thailand's tiger temple. *Society & Animals* 21(3), 266–283. DOI: 10.1163/15685306-12341282.

Cohen, E. (2014) Bullfighting and tourism. *Tourism Analysis* 19(5), 545–556. DOI: 10.3727/108354214X14116690097738.

Cohen, E. (2018) The philosophical, ethical and theological groundings of tourism – an exploratory inquiry. *Journal of Ecotourism* 17(4), 359–382. DOI: 10.1080/14724049.2018.1522477.

Cohen, E. (2019) Posthumanism and tourism. *Tourism Review* 74(3), 416–427. DOI: 10.1108/tr-06-2018-0089.

Constitutional Council (2012) Decision no. 2012–271 QPC of September 21, 2012. Available at: https://www.conseil-constitutionnel.fr/en/decision/2012/2012271QPC.htm (accessed 10 March 2022).

Deccan Chronicle (2017) Over 1 lakh throng Chennai's Marina Beach to protest jallikattu ban. *Deccan Chronicle*. Available at: https://www.deccanchronicle.com/nation/current-affairs/200117/over-1-lakh-throng-to-chennais-marina-beach-to-protest-jallikattu-ban.html (accessed 4 March 2022).

Dredge, D. and Jamal, T. (2015) Progress in tourism planning and policy: a post-structural perspective on knowledge production. *Tourism Management* 51, 285–297. DOI: 10.1016/j.tourman.2015.06.002.

Farooqui, A. (2001) *Early Social Formations*. Manak Publications PVT, New Delhi.

Fennell, D.A. (2012) Tourism and animal rights. *Tourism Recreation Research* 37(2), 157–166. DOI: 10.1080/02508281.2012.11081700.

Fennell, D.A. (2013) Tourism and animal welfare. *Tourism Recreation Research* 38(3), 325–340. DOI: 10.1080/02508281.2013.11081757.

Gosling, D.L. (2001) *Religion and Ecology in India and South East Asia*. Routledge, London and New York.

Guia, J. (2020) Conceptualizing justice tourism and the promise of posthumanism. *Journal of Sustainable Tourism* 29(2–3), 503–520. DOI: 10.1080/09669582.2020.1771347.

Hart, P. (2014) Dancing with bulls? An exploration of Indus iconography showing dynamic interaction between humans and bulls. Available at: https://www.academia.edu/2142088/Dancing_with_Bulls_An_exploration_of_Indus_iconography_showing_dynamic_interaction_between_humans_and_bulls (accessed 2 March 2022).

Hoyt, G. (2017) Fighting against bullfighting: tackling Spain's bloody tradition. Available at: https://digitalcommons.salve.edu/pell_theses/117/ (accessed 27 February 2022).

Hughes, P. (2001) Animals, values and tourism — structural shifts in UK dolphin tourism provision. *Tourism Management* 22(4), 321–329. DOI: 10.1016/S0261-5177(00)00070-4.

Humane Society International (2020) Bullfighting. Available at: https://www.hsi.org/news-media/bullfighting/ (accessed 1 March 2022).

Imranullah, M. (2017) Dalit artist was among earliest victims of jallikattu. Available at: https://www.thehindu.com/news/national/tamil-nadu/Dalit-artist-was-among-earliest-victims-of-jallikattu/article17094991.ece (accessed 3 March 2022).

Jayashree, B., Arul, A. and Yasmin, I. (2019) The voices of culture, conservation and the media event around bullfight jallikattu in Tamil Nadu, India. *Journal of Media and Communication Studies* 11(3), 20–30. DOI: 10.5897/JMCS2018.0644.

Joshi, J.P. and Parpola, A. (1987) Ignca.gov.in. Available at: https://ignca.gov.in/Asi_data/76884.pdf (accessed 2 March 2022).

Kalaiyarasan, A. (2017) Politics of jallikattu. Available at: https://www.jstor.org/stable/44166164 (accessed 5 November 2023).

Kumar, R. (2017) Tradition, culture or cruelty? Available at: https://www.thestatesman.com/features/tradition-culture-or-cruelty-1485982657.html (accessed 2 March 2022).

Kutubuddin, T.K. (2017) Know about the 7 disputed animal sports in India which are similar to jallikattu. Available at: https://thelogicalindian.com/story-feed/sports/animal-sports-india/ (accessed 2 March 2023).

Lal, A. (2017) Listen closely to the voices at Marina for the story beyond jallikattu, 'Tamil pride.' Available at: https://indianexpress.com/article/explained/listen-closely-to-the-voices-at-marina-for-the-story-beyond-jallikattu-tamil-pride-4486943/ (accessed 4 March 2022).

López-López, Á. and Quintero Venegas, G.J. (2021) Animal dark tourism in Mexico: bulls performing their own slaughter. In: *Exploring Non-Human Work in Tourism*. De Gruyter Oldenbourg, Berlin, pp. 69–82. DOI: 10.1515/9783110664058.

Markwell, K. (2015) Birds, beasts and tourists: human–animal relationships in tourism. In: *Animals and Tourism*. Channel View Publications, Bristol, UK, pp. 1–24. DOI: 10.21832/9781845415051.

Maudet, J.B. (2006) The bull marks his territory (Bullfighting festivities and territorial identities in Southwestern Europe and Latin America). *Annales de Géographie* 650(4), 361–387. DOI: 10.3917/ag.650.0361.

Moorhouse, T., D'Cruze, N.C. and Macdonald, D.W. (2016) Unethical use of wildlife in tourism: what's the problem, who is responsible, and what can be done? *Journal of Sustainable Tourism* 25(4), 505–516. DOI: 10.1080/09669582.2016.1223087.

Moorthy, S.N. (2018) *Jallikattu: New symbol of Tamil angst: An Analytical Narrative*. Vitasta, New Delhi.

Rajamannar, S. (2012) *Reading the Animal in the Literature of the British Raj*. Palgrave Macmillan, New York. DOI: 10.1057/9781137011077.

Ramesh, J. (2018) *Indira Gandhi: A life in nature*. Simon & Schuster India, Noida, India.

Rangarajan, M. (1998) The Raj and the natural world: the war against 'dangerous beasts' in colonial India. *Studies in History* 14(2), 265–299. DOI: 10.1177/025764309801400206.

Ravishankar, S. (2016) Jallikattu ban: how the centre is attempting to give the Animal Welfare Board of India a bloody nose. Available at: https://scroll.in/article/805951/jallikattu-ban-how-the-centre-is-attempting-to-give-the-animal-welfare-board-of-india-a-bloody-nose (accessed 9 March 2022).

Regad, C. and Riot, C. (2019) The Toulon declaration (English version) – univ-tln.fr, Université de Toulon. Available at: https://www.univ-tln.fr/IMG/pdf/the_toulon_declaration_english_version_.pdf (accessed 22 January 2023).

Regan, T. (2004) *The Case for Animal Rights.* Updated edn. University of California Press, Berkeley, California.

Reuters (2015) *Denmark Bans Bestiality in Move Against Animal Sex Tourism*. Reuters. Available at: https://www.reuters.com/article/us-denmark-bestiality-idUSKBN0NC1YM20150421 (accessed 21 January 2023).

Sarukkai, S. (2017) When the unelected set the agenda. Available at: https://www.thehindu.com/opinion/lead/when-the-unelected-set-the-agenda/article17342340.ece (accessed 4 March 2022).

Sen, S.N. (1999) *Ancient Indian History and Civilization*. New Age International, New Delhi.

Shani, A. and Pizam, A. (2008) Towards an ethical framework for animal-based attractions. *International Journal of Contemporary Hospitality Management* 20(6), 679–693. DOI: 10.1108/09596110810892236.

Shapland, A. (2013) Jumping to conclusions: bull-leaping in Minoan Crete. *Society & Animals* 21(2), 194–207. DOI: 10.1163/15685306-12341302.

Sharma, G. and Singh, S. (2015) Regulating India's blood-sport: an examination of the Indian Supreme Court's decision in Animal Welfare Board of India V. A. Nagaraja. *Jindal Global Law Review* 6(1), 113–122. DOI: 10.1007/s41020-015-0008-1.

Sheppard, V.A. and Fennell, D.A. (2019) Progress in tourism public sector policy: toward an ethic for non-human animals. *Tourism Management* 73, 134–142. DOI: 10.1016/j.tourman.2018.11.017.

Smith, S.R. (2012) Balancing the bull: Smit NO v His Majesty King Goodwill Zwelithini Kabhekuzulu. DOI: 10.10520/EJC153184.

Suroor, H. (2012) Jallikattu: Irish group calls for tourism boycott. Available at: https://www.thehindu.com/news/international//article60570338.ece (accessed 3 March 2022).

Tamil Nadu Tourism (n.d.) Chennai,Tamil Nadu, India: TTDC. Available at: https://www.ttdconline.com/jallikattu.html (accessed 8 November 2023).

Thakur, D. (2018) Resolving the cultural right-animal right conflict in India: analyzing Article 29(1) through the paradigm of jallikattu. *Commonwealth Law Bulletin* 44(3), 363–380. DOI: 10.1080/03050718.2019.1635510.

The Economic Times (2017) *Jallikattu: Marina Beach teems with protesters, stir peak*. Available at: https://economictimes.indiatimes.com/news/politics-and-nation/jallikattu-marina-beach-teems-with-protesters-stir-peaks/articleshow/56686047.cms (accessed 4 March 2022).

Thurston, E. and Rangachari, K. (1909) Castes and tribes of Southern India. Available at: https://indianculture.gov.in/rarebooks/castes-and-tribes-southern-india-vol-iii-0 (accessed 2 March 2022).

UK Court (2007) *Hunting Act*. UK High Court, United Kingdom. Available at: https://www.animallaw.info/case/r-application-countryside-alliance-and-others-v-attorney-general-and-another (accessed 10 March 2022).

Vishnoi, A. (2016) Animal Welfare Board gives Prakash Javadekar dos & don'ts list on culling. Available at: https://economictimes.indiatimes.com/news/politics-and-nation/animal-welfare-board-gives-prakash-javadekar-dos-donts-list-on-culling/articleshow/52892264.cms (accessed 9 March 2022).

Vivekanandan, P. (2013) Breed keepers and best practices. Linking breeds, natural resources, indigenous knowledge and institutions. Edited by P. Vivekanandan. Sustainable-Agriculture & Environmental Voluntary Action (SEVA). Published in 2012, pp. 167. *Animal Genetic Resources/Ressources Génétiques Animales/Recursos Genéticos Animales* 52, 169. DOI: 10.1017/S2078633613000180.

Wearing, S. and Jobberns, C. (2010) Ecotourism and the commodification of wildlife: animal welfare and the ethics of zoos. In: *Zoos and Tourism*. Channel View Publications, Bristol, UK, pp. 47–58. DOI: 10.21832/9781845411657.

Winter, C. (2020) A review of research into animal ethics in tourism: launching the annals of tourism research curated collection on animal ethics in tourism. *Annals of Tourism Research* 84, 102989. DOI: 10.1016/j.annals.2020.102989.

7 Bullfights, Tourism and Intangible Cultural Heritage in Mexico: A Dissonant Heritage

Georgina Flores Mercado*
Universidad Nacional Autónoma de México, Mexico City, Mexico

Abstract

This chapter analyses the relation between tourism, intangible cultural heritage and human–animal relationships in Mexico. In particular, it centres on bullfights and their heritage-making process and their legitimatization through the tourism market from a discursive perspective. Different official decrees and animal rights laws were used to make this discursive analysis. The results aim to establish that bullfights are a dissonant heritage in Mexico. There is a social dissonance between animal ethics and intangible cultural heritage discourses.

Introduction

Tourism is one of the most notable economic phenomena of the 20th and 21st centuries. However, tourism is not only an economic process but also a cultural one. It produces more symbols than objects and consumption is above all cultural consumption, where contemporary tourists consume musical, theatrical or artistic spectacles, exhibitions, fairs, fashions, cuisine, handcrafts, nightlife, sports events, etc. (Cohen, 2005; López and Marín, 2010).

This expansion of tourism has been facilitated by a joining up with other sectors, both social and cultural. Without a doubt, the cultural sector, using heritage, has been among the most successful. Heritage tourism is one of the branches of tourism that has employed millions of people directly and indirectly since the end of the sixties (Smith, 2006). Cultural heritage can transform a place into a tourist destination (Kirshenblatt-Gimblett, 1998), or those cultural practices valued as intangible heritage are used as bait to lure possible tourists. People themselves are also transformed into attractions to be consumed through festivals, museums and spectacles where they can be photographed, filmed and reified (Kirshenblatt-Gimblett, 1998). This also extends to animals.

The tourist industry has intensified and facilitated proximity and access to animals to facilitate tourist experiences (Winter, 2020). The inclusion of social practices which imply human–animal interaction corresponds largely to this semantic of global tourism. They have been used in a wide variety of settings: safaris, whale watching, sea lion watching, bird watching, trekking through habitats like forests and deserts, the massive promotion of culinary dishes which depend on a certain species, animals as a form of transport, cat cafes, spa destinations, etc. (Markwell, 2015; Rickly and Kline, 2021).

*gflores@sociales.unam.mx

 Tourism, Heritage and Commodification of Non-human Animals (eds A. López López *et al.*)
DOI: 10.1079/9781800623309.0007

Nevertheless, as Yujie Zhu (2021) suggests, heritage tourism should not just be understood as a piece of economic machinery but above all as part of a political and ethical project linked to the power of dominant groups and the hopes of the marginalized.

In this way, cultural practices of impoverished populations, ecosystems and animals of the global south are integrated into the tourist order by national governments, who follow the guidelines of the large multinational corporations of tourism. In a similar way, UNESCO aims to generate sustainable economic development and strengthen local economies through the tourism of tangible cultural heritage and, more recently, intangible cultural heritage (Chaves *et al.*, 2014).

UNESCO Intangible Cultural Heritage:

> means the practices, representations, expressions, knowledge, skills – as well as the instruments, objects, artefacts and cultural spaces associated therewith – that communities, groups and, in some cases, individuals recognize as part of their cultural heritage. This intangible cultural heritage, transmitted from generation to generation, is constantly recreated by communities and groups in response to their environment, their interaction with nature and their history, and provides them with a sense of identity and continuity, thus promoting respect for cultural diversity and human creativity.
> (UNESCO, 2003, p. 2)

In 2003, the Convention for the Safeguarding of Intangible Cultural Heritage was adopted by UNESCO. Since then, many cultural practices – even cultural practices that involve non-humans like falconry, horse riding, camel racing and wine horses – have been included, mainly in its Representative List. Although the inclusion of cultural manifestations in the Representative List seeks to generate pride, affirm identity at a national level and make cultural diversity visible on an international level, the current global context has above all potentialized promotion on the circuits of the tourist market, mainly in the sector of so-called cultural tourism (Smith, 2006; Chaves *et al.*, 2014; Flores Mercado and Nava, 2016). Although this is not one of the official objectives of the List, it is one of the main motives for state parties to register a practice in the Representative List (Hafstein, 2009). In this way, achieving international recognition forms part of a marketing strategy by governments, mainly those in the global south, to attract international tourism to their countries (Timothy and Nyaupane, 2009).

It is important to mention the Convention, given its relevance and impact on cultural policies and tourist policies on a global level. We need to be able to visualize the anthropocentric character of this international instrument, where the effect is that on the Representative List we can find several cultural practices that clearly use violence against animals. For example, la Charrería, a mestizo Mexican cultural practice where horses and calves are harmed, was included in 2016 under the category of an equestrian tradition.

The inclusion of these cultural practices has aroused concern and polemic. For example, during the 13th Session of the Intergovernmental Committee held in Port Luis, Mauritius in 2018, concern was expressed about animal rights.

The officer in charge of the French Centre for Intangible Cultural Heritage, Lily Martinet, points out that the Convention provides little help to member states, nor does it offer any indication for conciliating the welfare of animals with the safeguarding of intangible cultural heritage (Martinet, 2021). Unfortunately, the anthropocentric vision of UNESCO, which exalts humanity and its cultural products, privileges the satisfaction and welfare of humans in preference to the rights of animals. The Convention sets out limits as regards what can or cannot be included on its lists, but this delimitation is marked by human rights and sustainable development.

The anthropocentric position of the Convention not only legitimizes violence against animals but also makes invisible the voices of those who question what occurs in Mexico and other countries. Mexico ranks third place for bullfights on a global scale and possesses the largest bullring in the world, but a clear majority of Mexicans are against bullfighting. According to a telephone survey carried out by the *Reforma* newspaper with some 500 people, 59% of Mexicans would vote in favour of banning bullfights, while some 31% would vote against such a measure. The survey also revealed that 73% of those surveyed considered bullfights to be an 'act of cruelty against animals', while 25% saw 'art and tradition' (Diario Reforma, 2019).

Nevertheless, in the race to obtain the heritage distinction, and with it greater visibility in the tourist market, groups belonging to political and economic sectors in several states have mobilized in Mexico to obtain official recognition of intangible cultural heritage for bullfights. These initiatives have occurred both at state and international level, like in 2018 when the Culture Secretary presented a candidacy file to UNESCO with the aim of getting bullfights onto the Representative List. The file, which was not approved, aimed to include on the list the 'Traditional festivities related to bull culture, the paradigm of the May Fair of Tenabo'. The town of Tenabo, in the state of Campeche, is situated on the route of the *Camino Real*, a historic route which forms part of the tourist attractions of the south-eastern region of the country.

At a national level, bullfights have been institutionalized as intangible cultural heritage in several states: the state governments of Aguascalientes, Colima, Guanajuato, Hidalgo, Michoacan, Queretaro, Tlaxcala and Zacatecas have recognized bullfights as intangible cultural heritage within their territories (*La Jornada*, 2022). These states have promoted different tourism competitive initiatives articulated to governmental programmes like *Pueblos Mágicos* (Magic Towns). On the other hand, we can find other states – Coahuila, Guerrero, Quintana Roo and Sonora – that have legislated for the banning of bullfights within their entities. Several municipalities in the country have joined this initiative (PAOT, 2017).

In this complicated panorama, bullfighting tourism is promoted in cities like Cancun or, until very recently, Mexico City. Bullfighting tourism has widened its horizons, largely as a response to the increasing discomfort caused by bullfights among the population. This semantic renovation seeks to take the emphasis away from the '*fiesta brava*' – as bullfights are also known – focusing instead on so-called bullfighting culture, designing a wider and more attractive tourist proposal which integrates historical, social, architectural and economic aspects (Ríos Elorza, 2020).

According to the above, it proves interesting to analyse in this chapter the heritagization of bullfights, the discursive process of making heritage by state governments in Mexico. As Ashworth and Tunbridge (1996) point out, cultural heritage implies the deliberate selection of objects, artefacts and cultural expressions from the past and the present to satisfy the demands of contemporary consumption. It is equally interesting to analyse the legislation which establishes the ban in other states. The analysis of these processes will be done using a critical discourse perspective (Fairclough, 1992; van Dijk, 1998). In order to reconstruct the different discourses, I analysed the decrees of six states where bullfights, and in some cases the *charrería*, have been declared intangible cultural heritage and where tourist activity is relevant. The decrees belong to the states of Aguascalientes, Colima, Guanajuato, Hidalgo, Tlaxcala and Zacatecas.

Aguascalientes was the first state to grant state heritage recognition to bullfighting in 2011. Since then, other decrees have followed. These decrees for the heritagization of bullfighting are of variable length: between two and sixteen pages. Different names are used to refer to bullfights, including bullfights, *tauromaquia*, *fiesta brava*, *fiesta de toros* and *festejo taurino*. Nevertheless, the official name in all the decrees is the *fiesta de toros*, a category which makes the violence involved invisible and underlines the playful and hedonist character of bullfights, in order to attract tourists. It is important to mention that each of these decrees is based on the precepts of the 2003 Convention.

At the same time, I will analyse the animal rights legislation of the four states – Coahuila, Guerrero, Quintana Roo and Sonora – which have banned bullfighting. Sonora was the first to adopt this position in 2013, followed by Guerrero in 2014, Coahuila in 2015 and Quintana Roo in 2019, although it is important to point out that bullfights continue in Cancun, a city located in the latter state, and which holds certification from the World Tourism Organization. In most of these cases, state legislations regarding animal welfare and protection were passed or reformed, rather than publishing specific documents as occurred with the heritage decrees. Consequently, for the analysis I only consider those aspects which allowed me to follow the discursive threads of the animal ethics and posthumanist perspective.

For my analysis I will employ the concept of dissonant heritage proposed by Ashworth and Tunbridge (1996), the notion of discourse of Fairclough (1992) and the concept of ideologies of van Dijk (1998). I also use approaches from the posthumanist perspective of animal ethics (Fennell, 2015) and of abolitionism (Francione and Charlton, 2015). From these conceptual frameworks we can ask the following comprehensive and hermeneutic questions: What is the bullfights' heritage-making discursive process in the decrees? How are the tourism and cultural heritage discourses articulated with other discourses to sort out the dissonance? On the other hand, what kind of animal ethics discourses are used to argue for the prohibition? Do the legislations that have banned the bullfights have a posthumanist perspective?

Before beginning the development of our theoretical section, it is important to point out briefly that the studies in Mexico on bullfights and tourism have focused mainly on the analysis of the behaviour of the tourist market (Cabrera, 2022) or on their promotion strategies (Ríos Elorza, 2020). Among the few studies carried out from a posthumanist perspective we can find that of López and Quintero (2021), who have classified bullfights as dark tourism, as it puts in the centre of the analysis the death of the bull and considers it to be a macabre performance. Dark tourism is defined as the offering for tourism of places associated with death, but, according to the authors, studies of this type of tourism have been guided by an anthropocentric perspective and have not included examples which involve the death of animals (Quintero and López, 2018).

For the case we are interested in here, my analysis aims to contribute to the discursive understanding of heritage construction where the human–animal relation is involved, within the context of tourism in Mexico, and its dissonances. For Ashworth and Tunbridge (1996), dissonance is an intrinsic characteristic of all cultural heritage as it is inserted in a social system where selection is inevitable: 'all heritage is someone's heritage and therefore logically not someone else's' (1996, p. 21). The question which emerges immediately is: between what elements does dissonance occur? In this case, the dissonance is ethical, where heritage discourses, tourism discourses and others articulate to resolve, hide or dilute, through their rhetoric, the dissonance and the tension between animal rights and the right to culture.

The materials for analysis are different institutional heritage documents as well as legal documents which prohibit bullfights. My analysis does not extend to the discourses emitted by the anti-bullfighting movement, or the opinions which exist among different sectors of Mexican society. This may be considered to limit somewhat my analysis, yet the texts analysed are hugely relevant having been published by institutions, articulated with others like UNESCO, and as such occupy a hegemonic place in the process of creation and legitimization of what is or is not considered heritage and of its uses for example in tourism.

This analysis offers discursive evidence and reflections which could be used to refine argumentative tools against the heritagization and touristification of bullfights. As Ortiz-Millán points out: 'arguments are needed to rationally convince those who must make the final decision about a possible ban' (2014, p. 207). In this sense the battle for generating changes in the unequal human–animal relations in the area of tourism and heritage, is, in large part, determined in the discursive field.

Conceptual Framework for the Analysis of Bullfights as Dissonant Heritage

Cultural heritage should be considered not as an object, place or cultural expression, so much as an assembly of relations between people, things, institutions, corporations, governments and the environment, as well as ideological and epistemological structures which produce and mobilize meanings (Harrison, 2013). Likewise, we aim to recover the centrality of the discursive dimension in the construction of meanings related to cultural heritage (Smith, 2006) within this assembly of relations.

To think about the different meanings around a cultural practice, in this case bullfights, I take up the concept of dissonance heritage of Ashworth and Tunbridge (1996). These authors built their concept starting from the theory of cognitive dissonance of Festinger (1957), which establishes that a person can experience an inconsistency between their beliefs and a determined situation, either because this is new, or because it lacks logic, or due to cultural or moral distance, etc. Festinger calls this inconsistency dissonance, which being psychologically uncomfortable will motivate the person to try to reduce the dissonance and achieve consonance. Ashworth and Tunbridge suggest that dissonance in heritage involves a discordance or a lack of agreement and consistency and this lack of agreement is intrinsic to heritage. For both authors, the inequalities that exist in the processes of creation of heritage products are evident, given that different groups, due to their social, cultural or geographical condition, remain outside the process of heritage creation (Ashworth and Tunbridge, 1996).

To talk of dissonance in the area of heritage is to recognize the multi-voiced nature of heritage, as well as the inevitability of conflict (Smith, 2006). This plurivocality should be considered in the case of bullfights as cultural heritage, where different discourses articulate giving way to social ideologies.

The word ideology can designate almost anything, from a contemplative attitude to a set of ideas orientated towards action. From the indispensable environment in which individuals live, their relations with social structure to the ideas that legitimize a dominant political power (Žižek, 2003). For my analysis I will use the concept of ideology of van Dijk. This author uses this concept in his studies of racism in Europe (van Dijk, 1998). This author understands ideology as ideas, beliefs and discourses that sustain groups (social movements, businessmen, academics, etc.) and give meaning to their worlds and that provide a basis for social practice. Ideologies may have negative functions, such as hiding the interests of powerful groups to legitimize relations of domination, but can also have a positive function helping dominated or subordinated groups to reinforce solidarity, organize struggle and sustain opposition. Both in their positive and negative aspect, ideologies protect the interests and resources of groups. In general, ideologies emerge from struggles and conflict, and are therefore situated and defined through discursive and non-discursive practice, a *we* and a *them*, where the *we* takes on a positive value and the *them* a negative one (van Dijk, 1998).

Ideologies are constructed through discourses, which can be defined as groups of meanings that constitute objects and subjects, entities and social relations. We start from the idea that discourse is a social practice, as it not only represents the world but also signifies it, and by doing so constitutes the world (Fairclough, 1992). Parker establishes, from a radical constructionist point of view, that discourses allow us to see things that are not really there and that once an object has been elaborated in a discourse it is difficult not to refer to it as if it were real (Parker, 1992). That's the case with heritage.

To carry out discourse analysis and reconstruct the ideologies of bullfights it is important to trace the semantic content of the discourses that we are interested in identifying in institutional and legal texts. The first of them is the heritage discourse. This is characterized by underlining universality, national identity, monumentality, the artistic aesthetic and the notion of risk (Smith, 2006). Another, the tourism discourse, is articulated with economic processes of bullfights, like the generation of employment. For those texts that legislate the prohibition of bullfights, it will be important to analyse if they start out from a posthumanist position (Cohen, 2018) and identify what kind of ethical discourses regulate the banning of bullfights (Fennell, 2015). Posthumanist discourse is characterized by an erasure of the frontiers in the human vs animal, nature vs culture, subject vs object dichotomies, among others, and for rejecting human exceptionalism (Cohen, 2018). With regards to ethical discourses, we can mention welfarism, characterized by a concern for the life quality of animals and not so much with whether animals should or should not be used for human ends. The welfare of animals can be considered in three dimensions: natural living, physiology and feelings/mental behaviour. Animal rights is based on inherent value, which means that individuals exist as ends-in-themselves. Characteristics like consciousness, intentionality and sentience are emphasized.

Utilitarianism proposes as its central argument the idea that the moral value of an action is only determined by its utility in producing happiness or pleasure for the greatest number possible of sentient beings. Animals are sentient beings and therefore they can experience pain and hence suffer. Nevertheless, there is an acceptance of the idea that animals can be used, for example, in experiments, if these will result in benefits for wider social sectors (Fennell, 2015). Finally, the abolitionist discourse proposes the idea that all sentient beings, human or non-humans, have one right: the basic right not to be treated as the property of others (Francione and Charlton, 2015).

Bullfights: Historical, Social and Legal Frames

Erik Cohen points out that bullfights are a polysemic phenomenon, difficult to define with precision. Nevertheless, this difficulty of definition lies not so much in their forms of expression as in the meanings and feelings that they give rise to. The definitions we can find in a vast literature range from considering the bullfight as a ritual, a game, a fiesta, entertainment, a sport or a spectacle. This diversity of concepts, as Cohen (2014) points out, is not neutral, nor naïve, but implies an attitude – and I would add, a political and ethical position – towards this cultural practice.

The origins of this cultural practice can be traced back to ancient Mediterranean cultures, although the immediate precedent for the Hispanic tradition would be the Roman circus games (Ramírez Sánchez, 2020). In the Middle Ages the bullfighting spectacle was firmly consolidated in the Iberian Peninsula, characterized by the tumultuous persecution of the bull. Later, two different modalities appeared: the chivalrous and the popular bull traditions. It was in the former where the death of the bull was established not only as an essential moment of the practice, but also as the culmination of the festivity (Ramírez Sánchez, 2020). The variety of the bullfight with the bullfighter on foot became the dominant form in Spain during the 17th century (Cohen, 2014), but it was not until the 19th century when *tauromaquia* spectacles consolidated their current forms and limits (Maudet, 2010).

Transatlantic *tauromaquia* covers a wide range of practices, although all contain some degree of violence against the bull: American bull riding, the *toro coleado* (the bull is brought to the ground by pulling its tail) in Colombia and Venezuela, the *rodeo montubio* in Ecuador (a series of activities on horseback where the riders demonstrate their equestrian skills), the *vaquejada* (two cowboys pursue a bull, lead it towards a goal and then force it to fall) and *rodeo criollo* in Brazil, the Chilean rodeo and many others, as well as the great diversity of bullfights themselves, including the Spanish bullfight, the Portuguese bullfight, the Camargue bullfight, the Landas bullfight, the *recortadores* bullfight (the bullfighter encourages the bull to pass as closely as possible to his body) and others (Maudet, 2010). In Mexico we find the *jaripeo* (involving riding bulls bareback), the *charreada* (the fiesta of the charros), the *rejoneo* (a bullfight where the bullfighter is on horseback), the *rodeo*, the *toro de once* (which in some places marks the beginning of carnival), the *topa*, the *toro caballero*, the *morisma*, and others, but it is the bullfight which dominates the social imagination as regards *tauromaquia* (Ramírez Barreto, 2006).

In Mexico, as in other Spanish-speaking countries, bullfights formed part of Spanish colonization. There is little agreement as regards the date of its introduction, but some sources suggest that the practice was introduced during the early colonial period (Flores Hernández, 1986; Ríos Elorza, 2020). Bullfights were imbricated into the routines of ranches, as these produced domestic cattle for consumption, as well as fighting bulls for recreational purposes in the bullfights (Ríos Elorza, 2020).

Quintero and López point out that currently Mexico has 15 localities that are the most active in bullfighting tourism, and that visitors are predominantly concentrated in the bullrings of Mexico City and Cancun. Although Mexico does not have much specialization in bullfighting tourism, nor is it promoted with such intensity as occurs in Spain, tour operators have nevertheless incorporated a visit to bullrings in their urban tours (Quintero and López, 2018).

At the same time, bullfights have become increasingly less popular among citizens, especially in urban areas. Studies reveal a loss of interest in Spain, both among the national population as well as among foreign tourists. This has resulted in a dramatic drop in the number of bullfights occurring (Arbour, 2013). A change in values among younger generations can be explained in part by the strong accusations made by groups and civil organizations that defend animal rights, and that not only oppose the domination and lack of ethics during bullfights but also the violent treatment previously given to the bull (Cohen, 2014).

Although regulations and prohibitions of bullfights are nothing new, most interesting are the arguments and reasons given for their prohibition. For example, the *Concilio Provincial Hispalense* banned clergy from attending bullfights, while Pope Pío the Fifth banned bullfights in Rome because he viewed them as an example of pagan entertainment (Ramírez Sánchez, 2020).

Only recently has the violence used against bulls been considered as an argument for prohibition. In Spain, the governments of different Autonomous Communities have foreseen the regulation of bullfighting spectacles. However, these regulations have come into conflict with the legislations of the Spanish central government (García Rubio, 2021), as occurred when the Catalan Parliament passed a law abolishing bullfights within its territory for considering them as cruel spectacles where animals were publicly tortured (Ridao, 2017). In 2016, the Spanish central government declared this prohibitionist measure to be unconstitutional and considered bullfights to be protected by law, having been declared intangible cultural heritage by Spain in 2013 (Ridao, 2017).

In Mexico, tensions and disputes can also be traced across time. In 1867, for example, when the last imperial strongholds were falling and the republican regime was being reinstalled in the capital, newspapers of the period expressed voices calling for the complete suppression of bullfights. The argument stated that this prohibition was necessary as part of a true education for the population. Bullfights were banned for a short time in Mexico City. Some other localities followed this example, but the measure was reversed shortly after (Flores Hernández, 1986).

At this moment the legal processes for their prohibition are focused on animal rights. Recently, a judge issued the order to have them suspended in the Plaza Mexico bullring (*La Jornada*, 2022) while the National Supreme Court considered that bullfights and cockfighting could not be declared intangible cultural heritage by state governments (*El Universal*, 2022).

The Social Construction of a Dissonant Heritage: Ideologies and Discourses

As a result of the discourse analysis of heritage and legislative decrees, I have identified two main ideologies regarding bullfights. The first I have denominated the bullfighting heritage ideology, while I have called the ideology that bans the practice the animalist ideology. Now I will present the discourses that structure these ideologies, as well as the articulation with other discourses, that is to say their interdiscursivity. As we can see, each ideology aims to produce rhetoric with at least two objectives: first turn bullfights into cultural heritage and reduce social dissonance, and then make different interests (economic, environmental or cultural) coherent. The second objective emphasizes the sentience of animals and violence used by humans against animals, that is to say discursive dissonance is built.

Bullfighting heritage ideology

A central discourse in the bullfighting heritage ideology is the traditionalist discourse. The different decrees are nourished by historical information, with ideas about inheritance and ancestry, as well as an essentialist and collectivist notion of the practice. These arguments are consistently used in the promotion of tourism in Mexico. This discourse characterizes bullfights as a practice that has never suffered modifications:

> attending a bullfight means attending an event which by tradition continues to remain the same, as the essence of this social festivity was the same yesterday, as it is today, and as it still will be tomorrow, because the magical spell has the same intensity.
>
> (H. Congreso del Estado de Zacatecas, 2013, p. 7)

The traditionalist discourse works to sustain the practice through time. It gives an idea of continuity and permanence more than discontinuity and fragmentation.

Another relevant discourse is the nationalist discourse. This discourse places emphasis on Mexican identity, a result of a harmonious fusion of two cultures: Spanish and indigenous. This nationalist discourse asserts a unity through the mixing of cultures and considers symbolic references, like bullfighting, to be shared by all:

> The meeting of two worlds set off by the arrival of the Spanish in Aztec lands brought as a consequence that food, as well as diverse arts and activities that were subjected to a fusion that has been admired by those at home and abroad.
>
> (H. Congreso del Estado de Zacatecas, 2013, p. 1)

> The *Fiesta Charra* and the *Fiesta de Toros* are traditions and customs which are products of the cultural mix which left Guanajuato with the imprint of a society very different to either the Spanish or Mesoamerican ones.
>
> (H. Congreso del Estado de Guanajuato, 2013, p. 10)

In this way the nationalist discourse erases cultural differences, class differences or differences of any other kind in order to fuse cultural identities, Spanish and indigenous, into a sole entity: the popular soul. The final phrase of the Zacatecas decree sums up the idea: 'May *tauromaquia* continue to be the deepest expression of the Mexican popular soul!' (H. Congreso del estado de Zacatecas, 2013, p. 11). The playful perception of the bull's death is attributed to the Mexicans' character, to their idiosyncrasy, essentialist attributes which constitute the Mexican as impulsive: 'A fiesta rooted in the idiosyncrasy of Mexicans, with a playful vision of death' (H. Congreso del Estado de Aguascalientes, 2011, p. 13).

Bullfighting as a practice provides cohesion and a sense of identity. The continuity of the practice is the continuity of the community: 'to the extent that the bullfighting community is kept alive the fiesta will be maintained as a sign of identity. If this disappears, we will be close to disappearing as a community' (H. Congreso del Estado de Aguascalientes, 2011, p.13).

The discourse underlines the fact that there should be respect in the country for different ways of being. In this way we can reconstruct the discourse of cultural diversity, closely linked with the objectives of the Convention, which is used to legitimize and avoid the banning of the practice, as well as stating that bullfights should be accepted and recognized as cultural heritage: 'We cannot deny that this is the age of cultural diversity!' (H. Congreso del estado de Zacatecas, 2013, p. 7).

The architecture of bullrings is also incorporated within this process of heritage recognition of bullfights. Bullrings are temples of the practice, through which the bullfighting identity of the place is magnified, and national and local values that are supposedly shared are materialized. By considering them as cultural heritage, their conservation can be covered by heritage policies. Hence the decrees point out their historical past, building characteristics and aesthetics:

> *Tauromaquia* has been and is admired by the people of Zacatecas. A demonstration of this is the construction of the San Pedro bullring which was the cathedral of the *fiesta brava* in this part of the country.
>
> (H. Congreso del estado de Zacatecas, 2012, p. 5)

> The 'La Petatera' bullring, an authentic legacy for Colima, was built by families of Colima craftsmen, using techniques transmitted from generation to generation. They only used local materials from the region and applied knowledge making the bullring a flexible structure as well as efficient for an earthquake zone.
>
> (H. Congreso del Estado de Colima, 2016, p. 3)

Placing value on the bullrings in the decrees establishes a link between tangible heritage and intangible cultural heritage. In this way tourists can experience both 'authentic' cultural tradition and the architectural beauty of bullrings.

The link between the *fiesta brava* with cattle farming and cattle ranches is one of the most important discursive strategies for justifying its existence:

> The territory of Guanajuato has been, since before its birth as a Federal Entity, the place of important cattle ranches and uncountable traditions. We can see this in the thirty-two

> cattle ranches which exist today in the territory, creating an irreplaceable source of employment and economy.
>
> (H. Congreso del Estado de Guanajuato, 2013, p. 11)

The economistic discourse that gives meaning to the cattle ranches and their relationship to bullfighting closes off the possibility for other employment opportunities by naming them as 'an irreplaceable source of employment'. By affirming that cattle farming has *always* been there, it is represented as belonging to a mythical eternal time and not to a concrete historical moment, Spanish colonization:

> In the bull fiesta, Aguascalientes has always been a place for wild cattle, since the times of the Ranch of Cieneguilla, once the property of the Company of Jesus, the ranch of Venadero, the ranch of Peñuelas, just to name the oldest, up to the eighteen cattle ranches present today in the entity.
>
> (H. Congreso del estado de Aguascalientes, 2011, p. 12)

The economistic discourse occupies a central place in the meaning of the *fiesta brava*, being the symbolic cement that legitimizes the spectacle of cruelty: 'The bullfighting fiestas are also relevant in the economic sense. Both at a municipality level, as well as for numerous citizens, they constitute a very significant source of income' (H. Congreso del Estado de Colima, 2016, p. 3).

As we mentioned above, bullfights represent not merely a popular festivity, they are also a cultural industry. As such they are governed by capitalist precepts focused on the accumulation of profit. Within this logic, bullfights have joined with other commercial sectors like the powerful tourist sector:

> From an economic point of view three fundamental elements intervene: the cattle farms, the bullfighters, matadors [the bullfighters who actually deal the final death blow to the bull], *novilleros* [the bullfighters who fight with *novillos*, bulls younger than two or 3 years old] and *subalternos* [auxiliary members of the bullfighter's team], and the commercial companies which run the bull rings. Around these, and equally important, are the concessions of different services outside the actual bull ring, before, during and after the bullfight; the spending made by tourists and other citizens around the bullfight itself which contribute to the income of hotels, restaurants etc.
>
> (H. Congreso del Estado de Hidalgo, 2012, p. 1)

One of the effects of the economistic discourse is the competitive style of the decrees themselves. It is almost as if the state governments are taking part in a race where each is trying to stand out in one aspect or another: for the number of cattle ranches, the number of bullfights, the number of outstanding bullfighters, the quality of its bulls, the size of its bullrings, the age of the practice in the region, etc. The decrees seem to construct a kind of 'taurometre' which can be used to define which entity can be considered more 'bullfighterly'.

Other effects of the heritage discourse as well as strengthening articulation – interdiscursivity – with the tourist sector have also generated the creation of institutions focused on looking after bullfighting culture, as occurs in the state of Tlaxcala: 'The Tlaxcala Institute for Bullfighting Development will be commissioned to supervise measures to guarantee the viability of the intangible cultural heritage of the *Fiesta de Toros*' (H. Congreso del Estado de Tlaxcala, 2012, p. 3).

On the other hand, we can also find discourses that in some ways are opposites to or distant from the traditionalist discourse. Such is the case with the artistic discourse, a discourse which is generally linked to the avant-garde and to progressive ideas. Nevertheless, this is one of the discourses most drawn on in these decrees to legitimize the practice:

> Music, literature and painting have been the artistic expressions which have been nourished by the atmosphere, the essence and the endeavours of the *jolgorio charro* [*Charro* (cowboy) fun] and the *jarana taurine* [bullfighting commotion]. For that reason, invaluable works of art have been conceived from these events, whose artists, recognized in other parts, are moved by the inspiration of the finely patterned language of the bullfight and the glorified fortunes of the *Fiesta Charra*.
>
> (H. Congreso del Estado de Guanajuato, 2013, p. 12)

The artistic discourse is used to impregnate bullfights with artistic qualities to cover up or dissipate the central objective of the bullfight: the annihilation of the bull.

On the other hand, the decrees not only mention but also exalt the bullfighters, who are considered ambassadors of culture for their deeds during bullfights. Their grim performance in the bullring gives visibility to the state and to Mexico on an international level:

> As regards bullfighting, Guanajuato gave birth to one the greatest universal bullfighters. Rodolfo Gaona Jiménez, who practically modified the forms and ways of bullfighting, producing a new era in bullfighting which has lasted until today.
>
> (H. Congreso del Estado de Guanajuato, 2013, p. 13)

The bullfighter's identity is constructed through several different discourses, but those that stand out are the anthropocentric, the individualist and the rationalist. The bullfighter is the man who represents the human race, whose rationality and creativity impose themselves to dominate the savage world occupied by the bulls. The image of the bullfighter contrasts with the absence of the inexistent bull, which is never mentioned in the decrees. Only indirect allusions to the bull are made when cattle are mentioned. Value is only recognized in purely economic terms, reinforcing the hierarchy in the human–animal relation, where life, in this case the bull, is in the service of humans.

On the other hand, most of the decrees make little mention of bullfighting associations. When they are mentioned, as in the case of the decree of the state of Guanajuato, they serve to legitimize the process of heritagization, pointing out that it is organized citizens who demand the continuity of bullfighting, not tourism businesses and institutions: 'There is a wide sector of organized society which is actively in favour of the permanence of these traditions, manifested through this Declaration' (H. Congreso del Estado de Guanajuato, 2013, p. 14).

During the process of heritagization, there is frequently an attempt to identify the risks that the cultural practice confronts. The decrees identify globalization and sociocultural changes as factors of risk. However, some of them point out that the principal risk is posed by opposition groups: 'Nevertheless, with the increase in strong social currents which radically oppose the celebration of bullfights, today we face the urgency of protecting this tradition' (H. Congreso del Estado de Colima, 2016, p. 3).

The threat is considered to be present throughout the country, not only affecting bullfights themselves but also everything related to them.

The animalist ideology

As we have seen in the case of the bullfighting heritage ideology, in animalist ideology, the presence or use of certain discourses is necessary to give coherence and unity to this ideology and to achieve rhetorical goals. In general, ideologies that analyse ethics related to animals in tourism articulate discourses or part of them.

As regards state legislation in Mexico, we can point out that there are three main axes which structure the legislations: the notion of animals as sentient beings with rights, the eradication of mistreatment and cruelty, and the human–animal relationship. Although legislations integrate ethical discourses towards animals, principally animal rights, welfare or utilitarianism, none of them integrates the posthumanist discourse.

In the first axis, legislations define an animal as a living organism, non-human, sensitive, capable of movement by itself and of response to the stimulations of the environment. The legislations also form a typology of animals according to the functions they carry out in the human–animal relationship. Among these different classifications we can find abandoned animals, work animals, domestic animals, animals for company, exotic animals, wild animals, guide animals, potentially dangerous animals, animals for exhibition, for protection, for spectacles and for zootherapy. In this way, the human–animal relationship is signified, in the utilitarianism discourse, through an emphasis placed on the roles played by animals in the lives of humans.

Human beings are considered superior to animals, a hierarchy constituted due to the anthropocentric discourse that dominates legislations such as the Animal Welfare Law of Guerrero:

> Animals are creatures which, in relation to humans, occupy an inferior level on the evolutionary scale; this makes us responsible for their welfare. Having supremacy carries with it an obligation, a responsibility, to act as guardians of inferior species in intellectual terms.
>
> (Animal Welfare Law of the state of Guerrero, 2014, p. 3)

Although the different legislations construct different types of animals, these differences are erased in the sense that they all consider animals to be sentient beings. The capacity of animals for experiencing sensation is a characteristic that is highlighted and that the distinct ethical discourses share. An example can be found in Article 4 of the Animal Protection Law of Sonora:

> All animals, whatever their living conditions, will, as a result of the passing of the current Law, receive legal recognition as living sentient beings. As a consequence, measures will be adopted to avoid their suffering or being caused suffering and pain directly or indirectly by human beings.
>
> (Animal Protection Law for the State of Sonora, 2018, p. 4)

A common idea in the different legislations is that suffering is provoked by human beings in animals, and this can be physical and/or psychological, producing anxiety or fear. In this way, as an effect of the animal ethics discourses, these legislations ban all forms of abuse. In a similar way a discourse of violence constitutes animal abuse as a form of violence within society.

> Recognize the existing link between cruelty to animals, anti-social behaviour and interpersonal violence, and be able to establish and promote mechanisms and corrective actions to combat them.
>
> (Animal Welfare and Protection Law of the State of Quintana Roo, 2019, p. 1)

It is interesting to note how the violence discourse establishes parallels between animals and certain groups in society like children, women and the elderly, erasing in these cases the dichotomous boundaries raised by the Cartesian vision between humans and animals (Ingold, 1994). 'In families where there is violence, this is more frequently aimed at the weakest, which includes the elderly, women, children and pets' (Animal Welfare Law of the state of Guerrero, 2014, p. 3).

Conclusions

Tourism, particularly international tourism, has intensified the use of animals for recreation and tourist leisure (Winter, 2020). At the same time the expansion of tourism on a global scale has led to an articulation with other sectors, like the cultural one. Hence, we find that tourism offers museums, historic city centres or archaeological sites. But the populations of the global south also offer to the tourist market their traditional cultural practices, ecosystems and animals.

In this context international instruments like UNESCO's (2003) Convention are used as a way of gaining a position in the global market (Timothy and Nyaupane, 2009). At the same time the humanist and anthropocentric vision of UNESCO allows cultural practices that involve violence against animals to be considered intangible cultural heritage, legitimizing and prolonging their existence. In a similar way international recognition silences or makes invisible national legislation for banning bullfighting as well as the conflict and dissonance that exists as regards these practices in their places of origin (Ashworth and Tunbridge, 1996). For this reason, it is important to take notice of and to analyse these processes.

The discourse analysis of heritage decrees allows us to take note of interdiscursivity in the area of heritage, that is to say, the articulation of different discourses, through which there is an attempt to resolve the social dissonance towards bullfighting. Discourses reconstructed around bullfighting heritage ideology such as the traditionalist, the nationalist, the economistic, the artistic, the anthropocentric, the environmentalist and that of cultural diversity, are used to legitimize and socially sustain the practice of bullfighting, constituting it as cultural heritage. These discourses are used to obscure the inequalities and hierarchies of certain groups with economic and political power over other human groups as well as over non-human animals, in this case the bulls.

On the other hand, in the animalist ideology reflected in the legislations, I reconstructed principally three animal ethics discourses: the welfarist, animal rights and utilitarian. These discourses signify the animal subject as well as the dichotomous human–animal relationship. Despite these significant changes in Mexican legislation, an anthropocentric discourse prevails that maintains the hierarchy between humans and animals, where the former hold supremacy over the latter. Banning bullfights is legitimized not because cruelty towards animals is considered bad, but because it might lead humans to be cruel to other humans. The decision is based on human interests not on those of animals. This reinforces the hierarchical relationship between humans and animals (Ortiz-Millán, 2014).

To end this chapter, I consider that UNESCO and the World Tourism Organization should consider animal ethics in a more serious and emphatic manner within cultural policies. There is a need for a posthumanist gaze. Including in the Convention lists those practices that provoke debates and polarization is in the end a way of positioning in favour of one social sector, generally groups with economic and political power, against the rights of animals. It promotes a vision of the world that has been ethically questioned, as has occurred with *charrería* or bullfighting. What is more, UNESCO finds itself in contradiction with international legal frameworks, like the Universal Declaration of Animal Rights, approved by UNESCO itself. In the same way in which UNESCO has positioned itself in favour of human rights, it should also do so in favour of animal rights.

Acknowledgements

This chapter has been translated by Nicholas Matthew Risdell. A special thanks to those who reviewed and evaluated this text.

References

Animal Protection Law for the State of Sonora (2018) Ley de protección a los animales para el estado de Sonora. Number 45 section II of the official Gazette of the state of Sonora.

Animal Welfare and Protection Law of the State of Quintana Roo (2019) Ley de protección y bienestar animal del estado de Quintana Roo. New POE law decree 344.

Animal Welfare Law of the state of Guerrero (2014) Ley de bienestar animal del estado de Guerrero. CNDH Publication.

Arbour, C. (2013) An internal crisis is causing Spain's beloved bullfighting industry to bleed. *Quartz*, 8 August. Available at: https://qz.com/112234/an-internal-crisis-is-causing-spains-beloved-bullfighting-industry-to-bleed (accessed 30 January 2023).

Ashworth, G. and Tunbridge, J. (1996) *Dissonant Heritage: The Management of the Past as a Resource in Conflict*. Wiley, Chichester, UK.

Cabrera, L. (2022) El turismo de interior en México. *Desde el Sur* 14(3), e0042. DOI: 10.21142/DES-1403-2022-0042.

Chaves, M., Montenegro, M. and Zambrano, M. (eds) (2014) *El Valor del Patrimonio: Mercado, Políticas Culturales y Agenciamientos Sociales*. Instituto Colombiano de Antropología e Historia, Bogotá.

Cohen, E. (2005) Principales tendencias en el turismo contemporáneo. *Política y Sociedad* 42(1), 11–24.

Cohen, E. (2014) Bullfighting and tourism. *Tourism Analysis* 19(5), 545–556. DOI: 10.3727/108354214X14116690097738.

Cohen, E (2018) Posthumanism and tourism. *Tourism Review* 74(3), 416–427. DOI: 10.1108/TR-06-2018-0089.

Diario Reforma (2019) Dicen no a corridas de toros. *Diario Reforma*, 11 September. Available at: https://www.reforma.com/dicen-no-a-corridas-de-toros/gr/vi121198?ap=1 (accessed 30 January 2023).

El Universal (2022) Corte considera que corridas de toros y peleas de gallos no son patrimonio cultural. *El Universal*, 15 June. Available at: https://www.eluniversal.com.mx/nacion/suprema-corte-considera-que-corridas-de-toros-y-peleas-de-gallos-no-son-patrimonio-cultural-inmaterial (accessed 30 January 2023).

Fairclough, N. (1992) *Discourse and Social Change*. Polity Press, Cambridge, UK.
Fennell, D. (2015) The status of animal ethics research in tourism: a review of theory. In: Markwell, K. (ed.) *Animal and Tourism. Understanding Diverse Relationships*. Channel View Publications, UK, pp. 27–43. DOI: 10.21832/9781845415051.
Festinger, L. (1957) *A Theory of Cognitive Dissonance*. Stanford University Press, California. DOI: 10.1515/9781503620766.
Flores Hernández, B. (1986) *La Ciudad y la Fiesta. Tres Siglos y Medio de Tauromaquia en México*. INAH, México.
Flores Mercado, G. and Nava, F. (eds) (2016) *Identidades en Venta. Músicas Tradicionales y Turismo en México*. Instituto de Investigaciones Sociales, UNAM, México.
Francione, G.L. and Charlton, A.E. (2015) *Animal Rights. The Abolitionist Approach*. Exempla Press, Utah.
García Rubio, F. (2021) La tauromaquia patrimonio cultural inmaterial entre su protección y persecución. *Revista Aragonesa de Administración Pública* 57, 221–263.
Hafstein, V. (2009) Intangible heritage as a list: from masterpieces to representation. In: Smith, L. and Akagawa, N. (eds) *Intangible Heritage*. Routledge, Abingdon, UK, pp. 93–111.
Harrison, R. (2013) *Heritage. Critical Approaches*. Routledge, Abingdon, UK.
H. Congreso del Estado de Aguascalientes (2011) Decreto por el que se Declara a la Fiesta de Toros Patrimonio Cultural Inmaterial del Estado de Aguascalientes (Decree of the Fiesta de Toros as Intangible Cultural Heritage of Aguascalientes State). Periódico Oficial del Estado de Aguascalientes.
H. Congreso del Estado de Colima (2016) Decreto por el cual se Declaran las Fiestas Charrotaurinas de Villa de Álvarez como Patrimonio Cultural Intangible del Estado de Colima (Decree of the Fiestas Charrotaurinas of Villa de Alvarez as Intangible Cultural Heritage of Colima State). Periódico Oficial del Estado de Colima.
H. Congreso del Estado de Guanajuato (2013) Decreto por el que se Declara a la Fiesta Charra y Fiesta de Toros Patrimonio Cultural Intangible del Estado de Guanajuato (Decree of the Fiesta Charra and Fiesta de Toros as Intangible Cultural Heritage of Guanajuato State). Periódico Oficial del Estado de Guanajuato.
H. Congreso del Estado de Hidalgo (2012) Decreto de la Fiesta de Toros como Patrimonio Cultural Inmaterial del Estado de Hidalgo (Decree of the Fiesta de Toros as Intangible Cultural Heritage of Hidalgo State). Periódico Oficial del Estado de Hidalgo.
H. Congreso del Estado de Tlaxcala (2012) Decreto de la Fiesta de Toros como Patrimonio Cultural Inmaterial del Estado de Tlaxcala (Decree of the Fiesta de Toros as Intangible Cultural Heritage of Tlaxcala State. Periódico Oficial del Estado de Tlaxcala.
H. Congreso del Estado de Zacatecas (2013) Decreto de la Fiesta de Toros como Patrimonio Cultural Inmaterial del Estado de Zacatecas (Decree of the Fiesta de Toros as Intangible Cultural Heritage of Zacatecas State). Periódico Oficial del Estado de Zacatecas.
Ingold, T. (1994) *What is an Animal?* Routledge, London and New York.
Kirshenblatt-Gimblett, B. (1998) *Destination Culture. Tourism, Museums and Heritage*. University of California, California.
La Jornada (2022) Juez prohíbe de forma definitiva las corridas de toros en la Plaza México. *La Jornada*, 11 June. Available at: https://www.jornada.com.mx/notas/2022/06/11/cultura/juez-prohibe-de-forma-definitiva-las-corridas-de-toros-en-la-plaza-mexico/ (accessed 30 January 2023).
López, A. and Marín, G. (2010) Turismo, capitalismo y producción de lo exótico: una perspectiva crítica para el estudio de la mercantilización del espacio y la cultura. *Relaciones* 123(31), 219–258.
López, A. and Quintero, J. (2021) Animal dark tourism in Mexico: bulls performing their own slaughter. In: Rickly, J.M. and Kline, C. (eds) *Exploring Non-Human Work in Tourism. From Beasts of Burden to Animal Ambassadors*. CPI books GmbH, Boston, Massachusetts, pp. 69–81. DOI: 10.1515/9783110664058.
Markwell, K. (ed.) (2015) *Animal and Tourism. Understanding Diverse Relationships*. Channel View Publications, UK. DOI: 10.21832/9781845415051.
Martinet, L. (2021) Balancing animal rights and the safeguarding of the intangible cultural heritage. In: Cornu, M., Vaivade, A., Martinet, L. and Hance, C. (eds) *Intangible Cultural Heritage under National and International Law. Going Beyond the 2003 UNESCO Convention*. Edward Elgar, Cheltenham, UK, pp. 152–161. DOI: 10.4337/9781839100031.
Maudet, J.B. (2010) *Terres de Taureaux. Les Jeux Taurins de l'Europe a l'Amérique*. Casa de Velázquez, Madrid. DOI: 10.4000/books.cvz.7680.

Ortiz-Millán, G. (2014) Ética para matador. Savater, los toros y la ética. *Tópicos, Revista de Filosofía* 46, 205–236. DOI: 10.21555/top.v0i46.652.

PAOT (Procuraduría Ambiental y del Ordenamiento Territorial de la CDMX) (2017) Bienestar Animal en las corridas de toros (Animal Welfare in bullfighting). PAOT, México.

Parker, I. (1992) *Discourse Dynamics. Critical Analysis for Social and Individual Psychology*. Routledge, London.

Quintero, G. and López, A. (2018) Tauromaquia y turismo oscuro en México: las corridas de toros como prácticas no éticas. *Teoría y Praxis* 24, 197–228.

Ramírez Barreto, C. (2006) *De Humanos y Otros Animales*. Dríada, México.

Ramírez Sánchez, A. (2020) La tauromaquia: entre la tradición y el agravio a la vida animal. Ms Thesis. UNAM, México.

Rickly, J.M. and Kline, C. (eds) (2021) *Exploring Non-human Work in Tourism. From Beasts of Burden to Animal Ambassadors*. CPI books GmbH, Boston, Massachusetts. DOI: 10.1515/9783110664058.

Ridao, J. (2017) Bullfighting: amid public entertainment, cultural heritage and animal protection. *Revista Catalana de Dret Públic* 54, 171–184.

Ríos Elorza, S. (2020) Turismo taurino como estrategia de conservación para un patrimonio en riesgo: el caso de las ganaderías de toros de lidia en Tlaxcala, México. *PASOS. Revista de Turismo y Patrimonio Cultural* 18(2), 253–263. DOI: 10.25145/j.pasos.2020.18.017.

Smith, L. (2006) *Uses of Heritage*. Routledge, London and New York. DOI: 10.4324/9780203602263.

Timothy, D.J. and Nyaupane, G.P. (2009) *Cultural Heritage and Tourism in the Developing World: A Regional Perspective*. Routledge, Abingdon, UK. DOI: 10.4324/9780203877753.

UNESCO (2003) Convention for the Safeguarding of Intangible Cultural Heritage. UNESCO, Paris.

van Dijk, T. (1998) *Ideología. Una Aproximación Multidisciplinaria*. Gedisa, Barcelona, Spain.

Winter, C. (2020) A review of research into animal ethics in tourism: launching the annals of tourism research curated collection on animal ethics in tourism. *Annals of Tourism Research* 84, 102989. DOI: 10.1016/j.annals.2020.102989.

Zhu, Y. (2021) *Heritage Tourism*. Cambridge University Press, Cambirdge, UK. DOI: 10.1017/9781108914024.

Žižek, S. (2003) *Ideología. Un Mapa de la Cuestión*. Fondo de Cultura Económica, México.

8 Bullfighting at the San Marcos National Fair: Animal Cruelty behind Tourism and Heritage

Alejandro Morales Pérez* **and Brenda Martínez Velasco**
Universidad Nacional Autónoma de México, Mexico City, Mexico

Abstract
In Mexico, bullfighting is a contested practice. Animal rights advocates have denounced the violence and suffering inflicted on bulls at what they see as a pointless display of cruelty and animal abuse. Its defenders, on the other hand, justify it as part of culture and tradition and vie for its preservation. In Aguascalientes, bullfighting has been officially declared heritage, a move that seeks to establish bullfights as a tourist attraction (*tauroturismo*) while legally reducing bulls to the category of commodities for economic exploitation. This chapter approaches the bullfighting debate from a posthumanist perspective. It analyses the role bullfights play at the San Marcos National Fair and questions heritage declarations as a move that often benefits only a privileged minority.

Introduction

Every year, in the months of April and May, the city of Aguascalientes celebrates the San Marcos National Fair (SMNF), an event of national and international note (Gutiérrez, 2012) that includes bullfights in its menu. A sector of society regards bullfighting as a cultural practice embedded in the local identity, and on 17 October 2011 the state government proclaimed it part of the Intangible Cultural Heritage of Aguascalientes (Lozano, 2011).

In Mexico, bullfighting has an important cultural role in some of the places in which it is still practised – mostly in festive contexts. However, it is not free of controversy. This chapter addresses the existing debate around bullfighting as an element of the cultural heritage of Aguascalientes and the promotion of '*tauroturismo*' (that is, tourism associated with the so-called *fiesta brava* (Ollero and Gallurt, 2015) as a form of tourism. To approach this scenario, we begin by analysing bullfights in the context of the SMNF. We then present the unfavourable opinions by some attendants of the Fair (tourists, local visitors) and finally argue that only a minority of attendants have a strong cultural connection to bullfighting.

The official proclamation of bullfighting as heritage in Aguascalientes has been used as an argument to support the notion that it should be considered of public interest (Martínez, 2021). Some have proposed the appointment of a Bullfighting Commissioner in charge of documenting, protecting, propagating and revivifying the practice, so as to promote '*tauroturismo*'. Moreover, there are those who go beyond and push for the recognition of this

*Corresponding author: alejandro.mp@hotmail.co.jp

DOI: 10.1079/9781800623309.0008

title at the national level and ultimately by the UNESCO Intangible Heritage list. These efforts have met with resistance and criticism by both local inhabitants and visitors to the Fair – most notably, animal rights activists. Initiatives have been proposed to abolish bullfights and, in the short term, to prohibit the entrance of minors to the spectacle (Lozano, 2011).

Villaseñor and Zolla (2012) argue that the act of listing elements of heritage is tantamount to classifying culture hierarchically, and that the institutional actors in charge of doing so are influenced by economic and political interests. In a similar vein, Howard (2003) criticizes the general belief that heritage is made of only positive elements, and warns that privilege is often concealed behind it. He goes on to say that heritage is often nationalistic, exclusive, sexist or backward, and to this we could add that it may also have dire repercussions on the well-being of non-human animals, as happens with bullfighting, a practice rooted in speciesism.

Bullfights: Between Heritage and Animal Commodification

Bullfights are not in the UNESCO list of intangible cultural heritage. The proclamation by the state government specifies in one of its articles that the registration of the activity in the national and the UNESCO lists must be sought (Lozano, 2011). When talking about heritage, the following questions arise: Who creates it? Which stories does it tell, and which does it conceal? Who does it represent? Who does it benefit? What is its purpose? A number of answers have been proposed and widely discussed in the specialized literature.

McDowell (2008) argues that the construction of heritage is a selective task responding to particular interests that seek to politicize the past and the present. Graham *et al.* (2000) explain that heritage is a thing of the past that has been selected in the present for specific purposes, and that its origin is situated in the tastes and values of the 19th-century elites, which, under the cover of culture, established a Eurocentric hegemony.

As for the beneficiaries of heritage, Howard (2003) and Villaseñor and Zolla (2012) maintain that heritage declarations are biased and rarely serve as an instrument to actually safeguard the integrity of a practice, especially in areas like tourism. Gravari-Barbas (2018) resorts to the concept of globalization of heritage in order to understand the way in which heritage production processes tend to occur in the present time. For example, in a tourist environment, heritage is increasingly perceived as an element of distinction that allows a destination to compete in a globalized world and, to a lesser extent, as an identity reference.

The SMNF links tourism to heritage, and bullfights are considered an essential part of the festivities promoted by the state government and the board of the SMNF. Moreover, for Gravari-Barbas (2018), bullfights are expensive and exclusive events. Marvin, on the other hand, points out that bullfights today are less about community celebration than a festival: 'the *corrida* has become little more than an economically exploitable entertainment, something which for them implies both sterility and loss of the cultural significance of the *corrida* in this totality' (1988, p. 66).

For Cohen, a bullfight reaches its climax the moment 'at which a matador, fighting on foot (rather than mounted) kills a bull in an arena in the presence of a mixed domestic and sometimes foreign public' (2014, p. 545). Marvin mentions that, from the 18th century onwards, the professional acts of *toreros* on foot became 'the most complex and pre-eminent of the various types of celebration based on the man–bull contests' (1988, p. 52). Also, Cohen explains that, for lack of a better term, bullfights can be conceived of as contests of antagonist animals initiated by humans:

> These contests fall into two broad subclasses: animal-versus-animal contests, in which animals fight one another in contexts devised and managed by humans; and human-versus-animal contests, in which humans fight with animals.
>
> (2014, p. 545)

Marvin (1988) maintains that 'in order to understand the *corrida* or the art of *toreo*, it is essential to understand the animal which is at the centre of it' (p. 85), for it is the *torero* – a self-conscious being – and not the bull, who receives all the attention and is the object of preoccupation. This suggests that if bulls are valued in *corridas*, it is because 'through their interaction with them in this particular event, men assert and demonstrate certain culturally valued qualities about themselves as human beings' (p. 86). Therefore, and in tune with capitalist social relations, at the *fiesta brava* bulls are treated by tourism as commodities, used for spectacle, which propitiates a cognitive distancing between human and non-human animals.

The 2012 Cambridge Declaration on Consciousness states that some non-human animals – including mammals, to which bulls belong – have a neurological base that makes them aware of who they are and of their situation. This means that consciousness is not exclusive to humans (Low, 2012), and therefore the notion of cognitive distancing must be discarded. For (2012, p. 5), as social theories have moved into a 'postmodern' period that differentiates humans' subjective experiences, so too are they moving into a 'posthuman' period in which the notions that traditionally have separated humans from animals are challenged almost every day from academic quarters. In fact, animals are now being regarded in a completely different light within academia: they are no longer objects to be studied and categorized, but their experiential lives now count, as do our experiences with them.

The commodification of bulls is not limited to the arena. As Marvin (1988) indicates, the raising of bulls for fighting increased in the second half of the 18th century as a result of the rise in the number of *corridas*. The need to ensure a stable supply of fighting bulls led to their selective breeding, as wild specimens are hard to come by. However, the very fact that bulls for fighting are bred contradicts the notion that they are wild (Marvin, 1988).

Notwithstanding all this, the discussions around bullfighting are often limited to the spectacle. Not all bulls bred for the purpose end up in the arena to be slaughtered before the aficionados. But for every one that makes it there, many others have been sacrificed in the selection process. López and Quintero (2021) write that the breeding of fighting bulls is a years-long process that involves selecting only the most aggressive exemplars, even though aggressiveness is a trait alien to their nature.

In their first year of life, calves are separated, branded and confined by sex (López and Quintero, 2021). Branding with a hot iron is a widespread cruel practice to which the animal is subjected in order to establish that its body is somebody else's property (Fig. 8.1). In their second year, young bulls and cows are submitted to a resistance and bravery test called '*tienta*', which consists in accosting them and inflicting pain on them with the aim of choosing the most qualified individuals that are to be reproduced. They are destined for the herd, while the ones discarded are sent to the slaughterhouse. In its third year, a young bull, known as *novillo*, may be used for drills with *toreros* and even killed in the process. Those that are not will be sold and sent to the bullfighting ring the following year (Brei, 2021; López and Quintero, 2021).

Bulls for Fighting from the Humanist and Posthumanist Perspectives

The appreciation of fighting bulls in general corresponds with an anthropocentric perspective inscribed in humanism. Urbanik (2012) explains that this worldview is anchored in the Scientific Revolution and the Age of Enlightenment, both movements taking place in Europe during the 16th to 18th centuries. To the extent that social theory advances the knowledge of the different subjective experiences of humans, says Urbanik, it should be possible to effect a change towards a posthuman view that challenges the separation between human and non-human animals. In the same vein, Braidotti maintains that 'humanism has developed through the ages as a model of civilization which has made the idea of Europe coincide with the universalizing powers of reflexive reason' (2015, p. 21, translation by the authors). Gray explains that 'in general terms, humanism is the idea that the human animal is the receptacle of a universally unique value' (2014, p. 66, translation by the authors). This author emphasizes three aspects of humanism: (i) that 'the philosophers of

Fig. 8.1. Branding with a hot iron (known as '*herrajes*') to identify a fighting bull. (From Highsmith (2018) published under CC BY 2.0.)

ancient Greece believed that human beings were special because they had the ability to reason, which other animals lacked' (p. 66); (ii) that 'the human mind reflects the order of the Cosmos'; and (iii) 'the idea that history is a chronicle of the advance of humanity in which rationality is ever greater' (p. 67, translation by the authors).

On the other hand, Egan (2011) and Fennell (2012) consider that both speciesism and anthropocentrism directly descend from the great chain of being, an old universal hierarchy theory according to which humans are above all other beings. The term 'speciesism' was coined by Ryder in 1970 in a leaflet with the same title (Ryder, 2020) and defined as a type of moral discrimination based on the differences between species. Another interpretation is that of Horta, who indicates that 'speciecism is the unfavorable and unjustified consideration or treatment of those who are not classified as belonging to a certain species' (2020, p. 168, translation by the authors).

As for the origin of anthropocentrism, Lecaros contends that the essential event that allowed the development of a humanist view of the world, enthroning human will and his dominion over nature (and other animals), was:

> The infiltration over the centuries of Jewish and Christian ideas in the Greco-Roman philosophical tradition, especially those of stoicism, a doctrine that positioned man, on account of his reason, as the maximum end to which all other things in the world are directed.
>
> (2018, pp. 23–24)

Finally, Horta maintains that 'anthropocentrism must be considered a form of speciecism' (2020, p. 192), and Faria and Paez define anthropocentrism as 'the position according to which the interests of human beings should be favored over the interests of non-human entities' (2014, p. 96).

In contrast to the discourse used by the Government of Aguascalientes and the board of the SMNF to position bullfights as a cultural–touristic element and normalize the reification and commodification of bulls, several publications with a posthumanist standpoint showcase the suffering and wounds inflicted on bulls during the *corridas*. An important aspect to keep in mind is that not all posthumanism is antispeciesist. However, Ferrando (2019) maintains that post-anthropocentrism is one of the areas of philosophical posthumanism, which questions the superiority of humans over non-human animals. Having this in mind, in the next pages posthumanism will be dealt with from a post-anthropocentric perspective.

Posthumanism, a perspective that seeks to overcome anthropocentrism and speciesism in the study of the relations between humans and animals, strives for the recognition that the former are not superior to the latter, but rather that all animals that possess a central nervous system (Low, 2012) deserve equal treatment, which has led to a reconsideration of relations among species. Bolton (2014) suggests that the current conceptualization of what is human entails animal oppression and that the discussion on how to change this hierarchy must be directed at reviewing the division between 'humans' and 'animals'. Posthumanism offers a way to question humanism's dominant paradigm and, at the same time, a central mechanism to establish the rights of subaltern classes, including animals.

A forensic report by Cossío *et al.* (2017) exposes the suffering experienced by bulls during the three stages of a *corrida*: '*tercio de varas*', '*tercio de banderillas*' and '*tercio de la muleta*', three episodes in which three different ways of hurting the animals are displayed. During the first episode, a horse rider known as a 'picador' wounds the bull on at least two occasions with sharp instruments known as *puyas*. As these are thrust into the bull's body, 'the pyramid-like point cuts the skin, the subcutaneous tissue, and the rhomboid and trapezoid muscles that govern the extension or lengthening of the bull's neck and head' (Cossío *et al.*, 2017, p. 6, translation by the authors). In the second episode, three pairs of *banderillas* ('goads') are stuck into the bull's back. Similar in shape to harpoons, these instruments are designed to remain in place and shred the bull's muscles with every single movement, thus weakening him through the loss of blood (Cossío *et al.*, 2017). In the last of these acts, called '*tercio de la muleta*' or '*tercio de muerte*' (literally, 'death's third'), the bull is stabbed for several minutes and finally lunged through with a double-edged sword directed to the heart.

However, if the sword is not thrust at the right angle (which is very common), the animal won't die instantly and will suffer several such injuries (Zanardi, 2012; Ortiz-Millán, 2014), depending on the sword's trajectory. If the bull survives, a 10-cm-long dagger is thrust into the spinal cord in order to provoke respiratory arrest (Hoyt, 2017; Rivero, 2018). Though the damage inflicted on all three episodes is obvious, Cossío *et al.* (2017) emphasize that in the last episode, the bull's desperation and frustration before a threat he can't overcome is evident. Fig. 8.2 illustrates a dying bull after the three episodes just described.

Cossío *et al.* (2017) point out that the negative emotional states experienced by bulls bring them stress and provoke the secretion of adrenaline and noradrenaline, leading to hypertension, hyperthermia, hyperventilation and sweating. Additionally, cortisol is secreted 5 minutes after the stress stimuli. Neither adrenaline nor cortisol has analgesic functions, which means that the pain is not reduced. The authors conclude:

> Intentionally provoked events, such as wounds, pain, negative emotions, physiologic alterations, and death without loss of consciousness, are situations to be prevented and avoided in any activity with animals, regardless of their species or zootechnical end. Bullfights are therefore contrary to what are considered good management practices nationally and internationally; they go against the principles of animal welfare and are ethically questionable. (2017, p. 10, translation by the authors)

Methodology

To tackle our analysis, we opted for triangulation, a methodological strategy using two or more data sources to look at the same phenomenon (Decrop, 1999). This makes it easier to confirm information, reduce bias and lend credibility to the findings (Decrop, 1999; Cisterna, 2005; Betrián *et al.*, 2013; Gökçe and Özge, 2014). We conducted a survey of 211 questionnaires to tourists at the SMNF and supplemented these with 44 on-site semi-structured interviews with local inhabitants of Aguascalientes in April 2019 and February 2022. The sample

Fig. 8.2. A bull about to receive the death blow. (From Castelazo (2010a) published under CC BY-SA 4.0.)

size was decided during the process, as it was found that the answers were similar and repetitive. Additionally, in April 2019, two in-depth interviews were conducted, one with the representative of the Hispanic American Bullfighting Cultural Promotion (HABCP) and the other one with a representative of different organizations against bullfighting.

In anticipation of the possible cancellation of the 2022 edition (as had happened in 2020 and 2021 during the COVID-19 pandemic), informants for the survey were contacted through the SMNF's Facebook page (2022a). Likewise, bullfighting being a sensitive issue, a 'netnographic' procedure (obtaining written discourses from the study of cyberspace communities) was adopted. Today, the Internet is an ideal medium for data collection, especially in the social sciences. Among its advantages is the possibility of capturing the voluntary and uninhibited opinions of users behind the screen (Kozinets, 1998). The questions were oriented to finding out the motivations of tourists to attend the SMNF, as well as their opinion on the prohibition of bullfights at the Fair. These questions allowed for the identification of the relevance of bullfights in the context of the Fair.

As for the interviews of residents of Aguascalientes, they are considered a useful tool to obtain information (meanings, opinions, values) about a given socio-cultural group (Díaz-Bravo *et al.*, 2013). In the 44 interviews conducted, participants were asked their opinion on bullfighting as part of the city's heritage and whether bullfights are part of their cultural identity. Finally, in the case of the two in-depth interviews with key actors, the objective was to know their diverging perspectives, a positive feature of qualitative research (Bonilla-Castro and Rodríguez, 2005). These interviews evidenced the debate over the prohibition of bullfighting in Aguascalientes. The information obtained was codified and the data quantified, so as to identify patterns and draw generalizations in a deductive way. The topics were: (i) the relevance of bullfights as a tourist attraction, (ii) the perspective of the public on the practice of bullfighting and (iii) the nexus between bullfights and local culture.

Bullfights: A Decadent Spectacle of the San Marcos National Fair

In the previous sections of this chapter, the link between bullfights and heritage has been examined to show how it leads to the commodification of animals and their exposure to cruelty for the benefit of the tourism industry. This section presents part of the results that are linked to the commodification of bulls and the pain and suffering inflicted on them by unethical practices such as bullfighting. The discourse centres on proving that bullfighting, promoted as an essential tourist attraction of the San Marcos National Fair, is a decadent spectacle with a declining popularity and is considered offensive by a large part of the local population and tourists.

In Mexico, bullfighting as a tourist attraction exists mostly in cities with a Spanish heritage, such as Aguascalientes, famous for its San Marcos National Fair (López and Quintero, 2021). For Esparza (2007) and Gómez (2007), this celebration has successfully defined the city's character like no other element: it brings together popular merriment, traditional celebrations, cultural activities, a 'bullfighting serial' – considered to be the second most important in Mexico – and a fair that is perceived as a repository of the multiple identities of the community. In this respect, in an interview conducted in 2019, a representative of the HABCP declared: 'Aguascalientes is a very important reference right now, and I think that a San Marcos Fair without bulls is not a fair.'

The notion that bullfights are indispensable for the celebration of the Fair has been questioned by the results of a survey conducted in 2017 by Martínez (2021) among attendants of the SMNF. On the question of whether they consider bullfights an important tourist attraction, 62% of the women and 55% of the men said they have no interest in them. These figures are even more revealing when it comes to the pain and suffering of animals: 97% of participants said that animal suffering was the reason why they avoided bullfights and other spectacles based on animal abuse. These results indicate that a real paradigmatic change is taking place among the new generations, by which a relatively unimportant practice in the context of the Fair is perceived as a cruelty-ridden practice. The representative of organizations against bullfighting confirmed its decline in popularity: 'Bullfighting is in general decline. It attracts less and less people each time and there are fewer events. Some events have been cancelled for their low turnout and they have less support, especially among the younger generations.'

As for tourists, though they have a tendency to link the SMNF to bullfighting, this does not mean that they consider bullfighting to be the major attraction, nor that they approve of it. Visitors who participated in our survey were asked to name the image they had in mind when they thought of the Fair and the elements that they considered representative of it. Common among the answers were bullfights, concerts and consumption of alcohol, but there were also some that linked the Fair with violence and cruelty towards animals.

Though bullfights occurred in most of the answers, they were not always positively appraised. Only a small group of aficionados expressed interest in them, whereas the large majority of participants were either disinterested or indifferent, when not utterly against it. Many saw the Fair as a space that promotes animal abuse and linked bullfights to cruelty, sadism, torture and violence.

> Unfortunately, it [the Fair] is associated with unpleasant spectacles of violence towards animals, such as cockfights, bullfights, and *charreadas*. (Female, Aguascalientes)

> Sadism towards herbivores and birds, that is why I don't attend. Such sadistic practices are imported: what's national about it? (Female, Mexico City)

Tourists were also asked about the events they expected to attend. Fig. 8.3 illustrates the frequency of the answers. Bullfights come after musical events, and some participants, especially women, declared that they were interested in anything not having to do with animal exploitation. Puleo (2004) maintains that the public display of animal torture is a symbolic act in which anthropocentrism and androcentrism converge (both concepts being closely related to domination), and that if

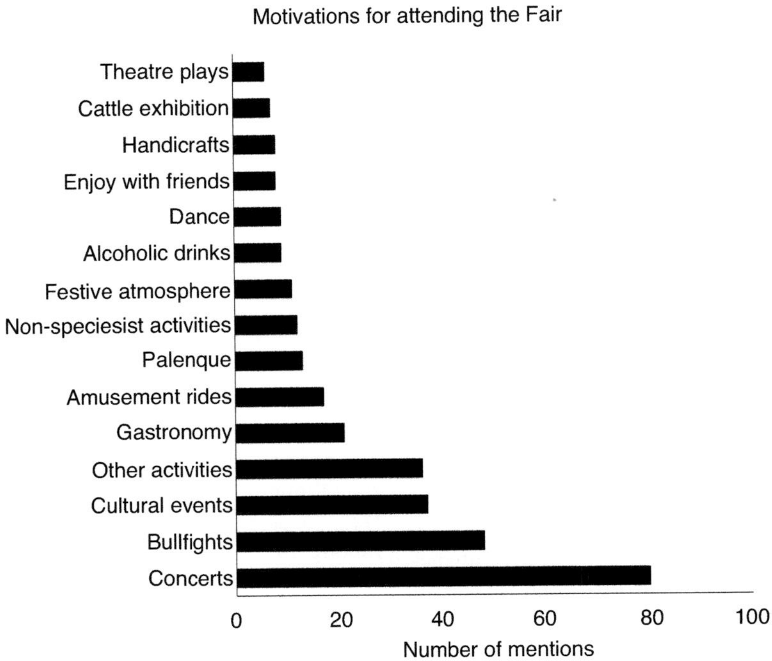

Fig. 8.3. Tourists' interests in SMNF activities. (Unpublished fieldwork data, collected in 2022.)

progress towards a more just society in which the domination, humiliation and elimination of otherness as identity-affirming practices is to be obtained, women's voices have a lot to say about bullfighting.

As can be seen in Fig. 8.3, bullfights are one of the main attractions of the San Marcos National Fair. However, their disapproval by tourists was manifest in our survey. Of 211 answers collected, 73% expressed no intention of attending a bullfight, and even those participants who admitted to having attended in the past expressed an adverse opinion of them:

> I only attended on one occasion, and I didn't like to see that people could enjoy watching how these animals were slowly killed. I don't understand. How can you enjoy watching something so sad? (Female, Aguascalientes)

> I went as a child, and never again in my adult life. I am against any form of abuse of any being. (Female, Mexico City)

Participants were also asked whether they thought that the Fair would continue to be attractive if bullfights were banned. Seventy percent of them said that it would continue to be as attractive or even more attractive without bullfights:

> Of course, it would be attractive; we don't need to teach the new generations to kill innocent beings. (Male, Aguascalientes)

> Yes, in fact it would be much more attractive if we tourists knew that there is no animal abuse. (Female, Mexico City)

The abovementioned seems to suggest that bullfights as a spectacle are important for just a minority of local residents and visitors – including, of course, the Fair's organizers. Many in Aguascalientes are of the opinion that sacrificing an animal for entertainment is wrong and that the tradition did not originate locally but came from elsewhere, so it shouldn't be considered as heritage. When local people were asked whether they considered bullfighting as representative of the cultural identity of Aguascalientes and its inhabitants, 24 participants of 31 responded affirmatively, alluding

Fig. 8.4. Spectators witnessing a bullfight in Aguascalientes. (From Castelazo (2010b) published under CC BY-SA 3.0 and GNU Free Documentation Licence.)

to the businesses linked to the theme (such as farms and bullrings) but especially to tourism in the context of the Fair (Fig. 8.4). The other participants said that bullfights were by no means representative of Aguascalientes, as they are a foreign custom:

> Look, to begin with, bullfights are a Spanish legacy. It didn't originate here as something traditional of Aguascalientes. (Male, 52)

> No, my point is that it is a Spanish tradition. (Female, 23)

Another question posed to local inhabitants was whether they personally identified with the bullfighting culture. Only seven participants responded affirmatively, and only four of them (an ex-worker of the bullring, a worker, the manager of a theme restaurant and a *banderillero*) said they attended bullfights; the rest, though they did consider themselves aficionados, said that they had attended very occasionally, that they had lost interest or that they watched bullfights on TV:

> I like *charreadas* [Mexican rodeo]. As for bullfights ... I am not so keen – why should I lie to you? I like watching them on TV. (Male, 39)

Behind the waning interest in bullfights among the inhabitants of Aguascalientes is a growing awareness of animal abuse. Apart from the *banderillero*, all participants answered affirmatively to the question of whether they were aware that bulls experienced pain and suffering in the bullring. Interviewees expressed this in their own words:

> I think that any harmful or aggressive act towards any animal involves suffering. They too have nerve terminals, even though they might not conceptualize it [suffering] as we do, but there is no doubt that some biological function is there, which brings them discomfort. (Male, 31)

> No, bulls are especially bred for fighting, we have had disagreements with the prohibition of bullfights, but in reality, bulls are reared for the purpose ... Yes, of course we have been accused of being assassins and such, many people don't like seeing an animal being tortured, but if you

> look closely, the animal comes out and puts a fight, so I don't understand why there is so much insistence on bullfights if cockfights are just as bloody and two animals kill each other, and if you go to the slaughterhouse, you can see how animals are sacrificed. (Male, 54)

Another aspect to explore was heritage. Despite the 2011 official declaration, when asked to name elements of their cultural heritage the majority of participants said they have no knowledge of it and only five of them, mostly aficionados, answered that bullfighting had been declared state heritage. These results lead to the conclusion that the majority of people of the city of Aguascalientes do not perceive bullfights as part of their cultural heritage and consider them entertainment of little interest. Howard (2003) points out that heritage is a concept linked to everything people want to protect and transmit, including their culture, regardless of whether or not this is featured in the tourist offer. Similarly, Jouault (2018) explains that a cultural element becomes heritage by social agreement, quite apart from official heritage lists. This seems to be the case in Aguascalientes, where bullfights are far from resembling the image that organizers and tourist agents are intent on projecting to the outside world.

Finally, although an undeniable link exists between the Fair and bullfighting, it would be misleading to conclude that the latter is the main event or the most awaited by visitors (see Fig. 8.3). Many equally or more attractive entertainment options exist, and many of them are free of charge. Bullfights are also not an indispensable activity in economic terms. An official report indicates that a total of US$420 million were collected over the 3 weeks of the 2022 SMNF (SMNF, 2022b). Though no mention is made of the money generated by bullfights (not surprisingly, given the secrecy that prevails when talking about numbers and bulls (Lorca, 2021), if, as mentioned in the report, 100,000 people attended the 14 bullfighting events and the entrance ticket to the ring costs an average of US$50, the money raised would be only US$5 million. Even allowing an extra US$10–15 million from the sales of food and drinks within the ring, this activity would be contributing a maximum of US$20 million, that is, less than 5% of the total. If, for some, bullfights are an indispensable feature of the Fair and the cultural heritage of Aguascalientes, the truth is that among the many, the widespread rejection of this practice is growing.

Conclusions

The city of Aguascalientes is the main Mexican tourist destination associated with bullfighting, partly on account of the inclusion of this activity in the San Marcos National Fair. Among the several factors that have made this city a reference in the matter are its Spanish heritage and the building of arenas and similar infrastructure. Fighting-bull farms abound in its surroundings, and thematic businesses related to the bullfighting culture thrive in its midst. Moreover, the fact that bullfighting has been declared cultural heritage in the state of Aguascalientes has strongly worked in favour of its inclusion in the Fair as a main attraction.

However, contrary to common perception, not all heritage nominations associated with tourism are positive, preservation-worthy cultural practices, especially when they involve the abuse of animals for human enjoyment. Bullfighting is a form of entertainment that can be approached from two opposed philosophical paradigms. From a speciesist humanist perspective, bulls are nothing more than commodities that can be traded in the name of 'tradition' and economic interest. However, from a posthumanist and post-anthropocentric perspective, bulls are not only sentient beings but are also self-conscious creatures with the ability to experience pain and suffering. This alone challenges the legitimacy of bullfights, independently of whether they are heritage or not.

The groups interested in the continuation of this practice are a wealthy minority with economic interests and enough political clout to influence decisions about the Fair. However, even if bullfights continue to be used as a symbol of the San Marcos National Fair, it does not mean they are the main attraction. The notion that bullfights are necessary for the celebration of the Fair has lost strength over the years, and in the opinion of many who have attended (both tourists and locals), this is a totally expendable spectacle, just one part of the Fair, whereas many others are cruelty-free and ethically unquestionable forms of entertainment.

Equally questionable is the cultural value of bullfights to the local inhabitants. As a practice that originated in Spain, many regard it as foreign and a substantial sector of the population of Aguascalientes disagrees with its promotion as an element of local culture. Not only is there little interest in this kind of spectacle (especially among women and young people), but also, as awareness about animal abuse grows, more people are totally against it.

Finally, the notion of heritage must be revisited: whose heritage and in whose benefit? Are harmful practices to be preserved only on account of their old age? Bullfighting as a tourist attraction benefits a small group of individuals, while a substantial part of society (both locals and tourists) rejects it altogether. A way out of this conundrum would be to adopt the postulates of post-anthropocentric posthumanism and begin by placing the essential interests of animals (preserving life and well-being) before non-essential human interests (such as entertainment and tourism). A paradigm change on the relationship between humans and animals is in order, so as to ensure the moral consideration of animals as beings entitled to dignified treatment, even if that means the eradication of age-old customs.

References

Betrián, E., Galitó, N., García, N., Jové, G. and Macarulla, M. (2013) La triangulación múltiple como estrategia metodológica. *Revista Iberoamericana Sobre Calidad, Eficacia y Cambio En Educación* 11(4), 5–24.

Bolton, B. (2014) Posthumanism and animal rights: rethinking 'The Human' rethinking the 'Self'. *Animal Studies Journal* 3(2), 48–56.

Bonilla-Castro, E. and Rodríguez, P. (2005) *Más Allá del Dilema de los Métodos. La Investigación en Ciencias Sociales*, 3rd edn. Grupo Editorial Norma, Bogota.

Braidotti, R. (2015) *Lo Posthumano*, 1st edn. Gedisa, Barcelona, Spain.

Brei, A. (2021) Torture and the Corrida. *Between the Species* 24(1), 22–38.

Castelazo, T. (2010a) Two seconds to death. Available at: https://commons.wikimedia.org/wiki/File:Two_seconds_to_death.jpg (accessed 29 January 2023).

Castelazo, T. (2010b) San Marcos bullfight. Available at: https://commons.wikimedia.org/wiki/File:San_marcos_bullfight_04.jpg (accessed 29 January 2023).

Cisterna, F. (2005) Categorización y triangulación como procesos de validación del conocimiento en investigación cualitativa. *Theoria* 14, 61–71.

Cohen, E. (2014) Bullfighting and tourism. *Tourism Analysis* 19(5), 545–556. DOI: 10.3727/108354214X14116690097738.

Cossío, A., Edwards, C. and Vanda, B. (2017) Dictamen forense sobre el dolor y sufrimiento de los toros durante la corrida, como evidencia de maltrato deliberado. Available at: http://www.paot.org.mx/micrositios/animales/pdf/toros/Dictamen_Forense.pdf (accessed 24 July 2021).

Decrop, A. (1999) Triangulation in qualitative tourism research. *Tourism Management* 20(1), 157–161. DOI: 10.1016/S0261-5177(98)00102-2.

Díaz-Bravo, L., Torruco-García, U., Martínez-Hernández, M. and Varela-Ruiz, M. (2013) La entrevista, recurso flexible y dinámico. *Investigación en Educación Médica* 2(7), 162–167. DOI: 10.1016/S2007-5057(13)72706-6.

Egan, G. (2011) Gaia and the great chain of being. In: Bruckner, L. and Brayton, D. (eds) *Ecocritical Shakespeare*. Routledge, London, pp. 57–69.

Esparza, V. (2007) Public amusements in the city of Aguascalientes during the porfiriato: in search of modernity. Master's thesis, El Colegio de San Luis, San Luis Potosí, Mexico.

Faria, C. and Paez, E. (2014) Anthropocentrism and speciesism: conceptual and normative issues. *Revista de Bioética y Derecho* (32), 95–103. DOI: 10.4321/S1886-58872014000300009.

Fennell, D.A. (2012) *Tourism and Animal Ethics: Contemporary Geographies of Leisure, Tourism and Mobility*, 1st edn. Routledge, Abingdon/New York.

Ferrando, F. (2019) *Philosophical Posthumanism*, 1st edn. Bloomsbury, London/New York. DOI: 10.5040/9781350059511.

Gökçe, O. and Özge, A. (2014) Film tourism triangulation of destinations. *Procedia – Social and Behavioral Sciences* 148(3), 625–633. DOI: 10.1016/j.sbspro.2014.07.090.

Gómez, J. (2007) *Historia de la Feria Nacional de San Marcos 1828-2006*, 1st edn. Gobierno Del Estado de Aguascalientes/Patronato Feria Nacional De San Marcos, Aguascalientes, Mexico.

Graham, B., Ashworth, G. and Tunbridge, J. (2000) *A Geography of Heritage*, 1st edn. Routledge, London/New York.

Gravari-Barbas, M. (2018) Tourism as a heritage producing machine. *Tourism Management Perspectives* 25, 173–176. DOI: 10.1016/j.tmp.2018.01.004.

Gray, J. (2014) *El Silencio de los Animales. Sobre el Progreso y Otros Mitos Modernos*, 2nd edn. Sexto Piso, Madrid.

Gutiérrez, R. (2012) Identidad visual de la Feria Nacional de San Marcos, análisis semiótico de los programas generales de 1980 a 2010. Master's thesis, UASLP, San Luis Potosi, Mexico.

Highsmith, C.M. (2018) Branding day at the Ladder Livestock Ranch. Available at: https://www.flickr.com/photos/vintage_illustration/45340965105 (accessed 29 January 2023).

Horta, O. (2020) ¿Qué es el especismo? *Devenires* 21(41), 163–198.

Howard, P. (2003) *Heritage: Management, Interpretation, Identity*, 1st edn. Continuum, London/ New York. DOI: 10.5040/9781350933941.

Hoyt, G. (2017) Fighting against bullfighting: tackling Spain's bloody tradition. Graduate thesis. Salve Regina University, Rhode Island.

Jouault, S. (2018) The Mayan milpa in Yucatán: a transition between patrimonialization and turistification? *Gremium* 5(11), 9–24. DOI: 10.56039/rgn10a03.

Kozinets, R.V. (1998) On netnography: initial reflections on consumer research investigations of cyberculture. *Advances in Consumer Research* 25(1), 366–371.

Lecaros, J.A. (2018) El puesto del hombre en la naturaleza: el problema del antropocentrismo. *Bioètica & Debat* 14(54), 22–25.

López, A. and Quintero, G. (2021) Animal dark tourism in Mexico: bulls performing their own slaughter. In: Rickly, J.M. and Kline, C. (eds) *Exploring Non-Human Work in Tourism: From Beasts of Burden to Animal Ambassadors*. De Gruyter Oldenbourg, Berlin/Boston, pp. 69–84. DOI: 10.1515/9783110664058.

Lorca, A. (2021) Pandemia taurina (I): los sueldos de los toreros, desde los mínimos del convenio hasta el gratis total. Available at: https://elpais.com/cultura/2021-09-03/pandemia-taurina-i-los-sueldos-de-los-toreros-desde-los-minimos-del-convenio-hasta-el-gratis-total.html (accessed 29 January 2023).

Low, P. (2012) The Cambridge Declaration on Consciousness. Available at: http://fcmconference.org/img/CambridgeDeclarationOnConsciousness.pdf (accessed 29 January 2023).

Lozano, C. (2011) Decreto por el que se declara a La fiesta de toros patrimonio cultural Inmaterial del estado de Aguascalientes. *Periódico Oficial Del Estado de Aguascalientes* 42, 13–14.

Martínez, B. (2021) El tauroturismo en la ciudad de Aguascalientes: la mercantilización de los toros de lidia bajo el discurso del patrimonio cultural. Graduate thesis, UNAM, Mexico City.

Marvin, G. (1988) *Bullfight*, 1st edn. Basil Blackwell, Oxford/New York.

McDowell, S. (2008) Heritage, memory and identity. In: Graham, B. and Howard, P. (eds) *The Routledge Research Companion to Heritage and Identity*. Ashgate, Hampshire/Burlington, pp. 37–53. DOI: 10.4324/9781315613031.

Ollero, A.C. and Gallurt, J. (2015) Nuevo concepto del turismo taurino. *International Journal of Scientific Management and Tourism* 4, 141–150.

Ortiz-Millán, G. (2014) Ética para matador. Savater, los toros y la ética. *Tópicos, Revista de Filosofía* 46, 205. DOI: 10.21555/top.v0i46.652.

Puleo, A. (2004) Feminismo y tauromaquia. *El Viejo Topo* 195–196, 72–77.

Rivero, P. (2018) En buena lid. In: Rivero, P. (ed.) *Zooética: Una Mirada Filosófica a Los Animales*. Fondo de Cultura Económica, Mexico City, pp. 56–67.

Ryder, R. (2020) Especismo. Available at: https://www.nipea.info/philosophia-naturalis/especismoderichard-ryder/ (accessed 29 January 2023).

SMNF (2022a) Official page of the Feria Nacional de San Marcos, Facebook. Available at: https://www.facebook.com/ferianacionaldesanmarcosoficial (accessed 29 January 2023).

SMNF (2022b) Official page of the Feria Nacional de San Marcos, Facebook. Available at: https://www.facebook.com/ferianacionaldesanmarcosoficial/videos/informe-de-resultados-fnsm2022/448618740429450/ (accessed 29 January 2023).
Urbanik, J. (2012) *Placing Animals: An Introduction to the Geography of Human-Animal Relations*, 1st edn. Rowman & Littlefield Publishers, Lanham, Maryland.
Villaseñor, I. and Zolla, E. (2012) Del patrimonio cultural inmaterial o la patrimonialización de la cultura. *Cultura y Representaciones Sociales* 6(12), 75–101.
Zanardi, T. (2012) National heroics: bullfighters, machismo and the cult of celebrity. *Journal for Eighteenth-Century Studies* 35(2), 199–221. DOI: 10.1111/j.1754-0208.2012.00497.x.

9 Restoring Relationships of Respect with Eagle Relatives: Tribal-led Eagle Aviary Tours in Oklahoma Indian Country

Bobbie Chew Bigby*
University of Waterloo, Canada and University of Notre Dame Australia, Broome, Australia

---

These eagle relatives are our prayer warriors. We now must return the favour that they have done for us humans. – Jennifer Randell, Citizen Potawatomi Nation Aviary Manager

Abstract

This chapter takes a focused look at the experience of Tribal-led Eagle Aviary tours in the US state of Oklahoma and specifically among the Citizen Potawatomi Nation (CPN). The CPN Eagle Aviary was opened in 2012 to provide care for injured eagles that cannot be released to the wild, along with providing tours and educational experiences to Tribal members and the wider public. Using a critical Indigenist approach and Indigenous storywork methodology, this chapter provides a snapshot of the Aviary tour experience to reveal the layers of Potawatomi culture, belief and practice that are shared in conjunction with visitors' experiences with eagles. This chapter asserts that while a posthumanist theoretical approach can be seen to resonate with several facets of the tour, an Indigenous resurgence framework is more strongly suited to understanding the overall tour experience, as well as the Aviary's objective of helping visitors to (re)establish relationships with these birds that are understood as eagle relatives with extraordinary capacities.

Introduction

Through the eyes of many Tribal Nations in Oklahoma Indian Country, eagles represent far more than just majestic, beautiful birds of prey or simply a national symbol of the USA. Instead, eagles are revered animals, understood as bird relatives that have the unique gift of flying so high in the sky that they serve as messengers between the Creator and humans on earth (Citizen Potawatomi Nation Cultural Heritage Center, n.d.). Because of these deeply held values and respect for eagles, these birds are invoked in numerous aspects of Tribal cultures, including in stories, prayers, songs, clan relationships, dances and also traditional clothing regalia.

Responding to the number of injured eagles and birds requiring long-term care (Iowa Tribe of Oklahoma, n.d.), several programmes led by Tribal communities have sprung up in recent years for these birds that can no longer survive in the wild. Across Oklahoma, three Tribally-led programmes exist, including aviaries managed by the Iowa, Citizen Potawatomi and Comanche Nations, all of which carry the designated US Fish & Wildlife Service Native American Eagle Aviary

*bbigby@uwaterloo.ca; bobbiechewbigby@gmail.com; bobbie.bigby@my.nd.edu.au

DOI: 10.1079/9781800623309.0009

Permit ('3-200-78: Native American Tribal Eagle Aviary'; US Fish & Wildlife Service, nd a). Primary goals of these Tribal aviaries are to rehabilitate and care for injured eagles. Beyond this goal of care, however, each of these Tribal programmes also has pursued other initiatives related specifically to educating and spreading awareness about eagles to the public through the form of tourism experiences. These educational tours and programmes allow visitors of all ages, ranging from students to Elders, to learn not only about eagle conservation and stewardship, but offer Native interpretations and understandings of these important birds through a culturally grounded lens. Rooted in their significant traditional relationships with eagles, these Tribally-led tour experiences can help Native communities to educate their own Tribal members, as well as the public, about the roles of these sacred animals and our responsibilities to them as stewards who traditionally maintain relationships with all living beings.

This chapter offers an overview of these Tribally-led eagle aviaries, with a specific focus on the tour experience provided by the Citizen Potawatomi Nation's Eagle Aviary, located in central Oklahoma. Applying a posthumanist lens to understanding these Tribally-led experiences, this chapter argues that these Indigenous-led eagle tourism initiatives do resonate with aspects of this theory given its rejection of anthropocentrism and human exceptionalism, exploitation and abuse in all human–animal interactions, including within the tourism space (Cohen, 2019). However, this study also asserts that the culturally grounded interpretations, practices and experiences of Potawatomi people that come through during the tour also push the posthumanist theoretical perspective to its limits. Building on literature that has examined the unique value of Indigenous interpretations in animal and ecotour experiences (Zeppel and Muloin, 2008a; Wright *et al.*, 2009; Zeppel, 2010; Walker and Moscardo, 2016), as well as Indigenous perspectives on the responsibilities and kinship that traditionally characterize human–animal interactions (Champagne, 2004; LaDuke, 2016), this research asserts that resurgence paradigms offer a stronger theoretical lens for understanding the Tribal-led aviary tour experience and the building of human–eagle relationships given that resurgence theories prioritize, uplift and work within Indigenous concepts and worldviews.

Literature Review

Animals, tourism, justice discourses and posthumanist perspectives

As noted by Kline, the research and documentation of animals within tourism activities is a field that has received increasingly more attention beginning in the 1980s, with a marked shift away from anthropocentric and species-centred framing and gradually towards lenses prioritizing justice, ethics and animal welfare (2022, p. 195). This has mirrored a similar and broader discourse shift more generally in the thinking and treatment of animals, away from notions of human exceptionalism and towards a stronger framework of animal rights and justice. This deeper ethical stance has been enabled not only through promoting compassion and concern for animals as living beings with their own set of rights and interests, but also through the advance in scientific research and explorations that have revealed the depths of animal self-awareness, sentience and capabilities (Bekoff and Sherman, 2004; Irvine, 2007; Edelman and Seth, 2009; Welsch, 2017).

Closely tied in with the discourses of animal rights and species justice is the field of posthumanist thought (Guia, 2021), a body of philosophy that rejects the idea of human exceptionalism, negates the human–animal divide and promotes instead a 'flat ontology' whereby humans and all other living beings are on a level, egalitarian playing field (Law, 2004; Cohen, 2019). Posthumanism's foray into tourism studies has only recently begun, starting with Cohen's article on 'Posthumanism and tourism' (2019) that attempts to explore this nexus in a conceptual way. Beyond advocating equal treatment for humans and animals within the tourism context, Cohen asserts that application of a posthumanist perspective or ethic would likely oppose many current forms of animal tourism that can be considered exploitative of animals through their commodification and/or enslavement. An application of this posthumanism would instead likely 'favor an entanglement of tourists with other humans, and with non-human animals in mutual, multi-sensual interaction, as a manifestation of the co-dependence between all living beings' (2019, p. 420). Cohen acknowledges that what exactly this 'mutual' entanglement might look like in tourism is unclear, yet offers as one possibility that this posthumanist direction of tourism might be able to fit within other hybridized modes of tourism, such as educational and academic tourism, that prioritize the learning that tourism activities can facilitate (2019, p. 421).

Another important stream of thought related to the involvement of animals in tourism are discourses based in relational theories. In countering long-held viewpoints by certain cultures that human–animal relationships are defined by dominance of humans over animal beings, relational theoretical frameworks offered by the likes of O'Neill (1997) and Donaldson and Kymlicka (2011) emphasize a clear agenda of human obligations towards animals as opposed to subjugation and dominance. These obligations or responsibilities can be understood as actions intended to protect, preserve and rehabilitate animals as a critical part of the natural world as opposed to exploiting them solely for human benefit.

Indigenous interpretations and Indigenous-facilitated experiences of animals

Within the context of Indigenous-led tourism and Indigenous interpretations that are offered during tourism experiences, scholars such as Walker and Moscardo (2016) have made the case for the importance of sharing of cultural values in impacting the perspectives and viewpoints of visitors (who are mostly non-Indigenous) on tours. Additionally, authors including Ruhanen and Whitford (2019) have drawn a clear connection between Indigenous tourism experiences and the links to cultural heritage and knowledge that are shared in these spaces. Yet situated within the field of animal experiences in tourism sits a small but important research focus subset that specifically examines Indigenous interpretations and/or Indigenous-facilitated experiences of animals in tourism settings. In terms of specifically linking tourism, animals and Indigenous cultural interpretations, one of the leading voices to explore this field is Heather Zeppel, with much of her writing devoted to understanding the varied intersections between Indigenous perspectives, tourism and wildlife (2006, 2008a,b; Zeppel *et al.*, 2003). Zeppel and Muloin (2008a, b) assert that unique cultural differences exist in the attitudes different human communities maintain towards animals. On the one hand, Indigenous perspectives are often centred around moralistic, utilitarian and aesthetic engagements with animals as reflected in spiritual beliefs, foodways and clothing; on the other hand, Zeppel and Muloin characterize non-Indigenous, mainstream attitudes towards animals as often based in more ecological and scientific understandings (2008a, p. 133). Regardless of the particular differences in attitudes and cultural understandings towards animals, however, their study ultimately asserts that Indigenous interpretations in wildlife tourism within the study's Australian context have 'the potential to expand and alter visitor understandings of Australian wildlife' (2008, p. 133).

Turning to the insights and voices of Indigenous tour facilitators themselves in the context of animal-based tourism experiences, the work of Wright *et al.* (2009) provides one of the most insightful and comprehensive accounts of the ways in which animal experiences in tourism can be facilitated by Indigenous peoples and cultural worldviews. Collaborating with Indigenous tour guides and owners of the Bawaka Cultural Experiences (BCE) tours in the Northern Territory of Australia, these authors collectively reflect on the ways that Indigenous Yolngu understandings of animals are shared through experiences in Yolngu Country and thus inspire visitors to re-think their own relationships with animals through an 'ontology of connection' that is modelled by Traditional Owners (Wright *et al.*, 2009, p. 521). For example, the BCE tour guide authors share numerous stories about the sophisticated communication and ceremonial relationships that community members maintain with crocodiles, who are regarded as sacred for the community.

The Bawaka community is said to have between 800 and 1000 different songs and stories about crocodiles and, during a tour, experiences, numerous anecdotes and stories about these deep relationships with crocodiles are often shared. The authors write that these tours are 'not simply a matter of teaching tourists about animals' (Wright *et al.*, 2009, p. 507); these BCE Indigenous-led tour experiences are instead better understood as opportunities 'to experience animals as related, sapient and connected with humans (and with rocks and trees and other things that are not seen as sapient within a western framework)' (Wright *et al.*, 2009, p. 507). It is through these opportunities created by Indigenous Bawaka sharing that many visitors' notions of animals and their relationships to animal beings can be fundamentally challenged, opening up innumerable opportunities for rethinking human 'ways of being in the world' (Wright *et al.*, 2009, p. 507).

Central to this Indigenous community's attitudes towards the animals that take part in the tour experience is the deep cultural conviction that 'animals, as well as plants, the land, the winds, the water and the elements, are sentient, sapient and have agency ... they have knowledge and wisdom as well as songs of their own' (Wright *et al.*, 2009, p. 516). Following this cultural logic, humans are compelled to regard these animal neighbours, along with all other beings making up the surrounding environment, with deep respect and consideration not simply because it is an ethical code but because these beings are agents in their own right with numerous skills that are different from and sometimes even superior to human abilities. What further distinguishes this Indigenous ethic of respect for animals from the ethical viewpoints articulated earlier, such as 'species justice' and notions of animal rights emanating from non-Indigenous discourses, is both the deep sense of interconnected relationships and responsibilities that Indigenous Yolgnu people have in relation to animals (along with other non-human beings in the environment) that are regarded as kin. As made clear in the writing of Wright *et al.*, a BCE tour experience allows visitors to glimpse not only the animals themselves and be exposed to the rich cultural stories connecting humans with the animals, but also to witness these active relationships in everyday life. Grounded in these Indigenous perspectives and relationships with animals within the context of tourism, it is helpful to turn towards an overarching Indigenous-led paradigm that can help to link the different elements of Indigenous peoples, cultural worldviews, animals and tourism engagements into a holistic and culturally informed conversation.

Resurgence as the alternative

Emerging from decolonial and Indigenist theoretical writings of scholars including Frantz Fanon, Vine Deloria Jr (Standing Rock Sioux), Jeff Corntassel (Cherokee Nation), Leanne Betasamosake Simpson (Michi Sagiig Anishnabeg) and Glen Coulthard (Yellowknives Dene), among others, the discourse of resurgence is an exciting field of interdisciplinary thought that is Indigenous-led. At its core, resurgence paradigms emphasize the importance for Indigenous peoples to re-establish and strengthen traditional, meaningful connections to cultural lifeways, lands and waters, as well as community, above any other goals. By prioritizing Indigenous understandings and concepts as the central terms of reference, as opposed to relying on mainstream concepts such as rights, justice and ethics that are heavily grounded in Western understandings, resurgence discourses aim to engage deeply with Indigenous ontologies, experiences and ways of being that have been devalued and intentionally undermined for so long through colonization. Scholar Gina Starblanket (Cree/Saulteaux) writes that 'Critical Indigenous and post-colonial scholars are increasingly interrogating the emancipatory potential of projects of self-determination that work within state sanctioned legal and political structures' (2018, p. 29). Instead of seeking recognition or conformity, many Indigenous people and communities are instead returning to practices, 'grounded in our ancestral values, places, knowledges and spirituality ... [ultimately towards] the healing and strengthening of our communities on our own terms' (Starblanket, 2018, p. 29).

Resurgence thinker Glen Coulthard uses the term 'grounded normativity' as a general description for the diverse, place-based ethical understandings on which Indigenous worldviews are based and often manifested through varied cultural forms, including song, story, ceremony and customary laws (2014). Undeniably at the heart of diverse Indigenous-grounded normativities are the stories, beliefs, thinking, experiences and relationships that Indigenous communities maintain and seek to renew with animal relatives. Life-long environmental activist Winona LaDuke (White Earth Anishnaabe) shares that from her own Anishnaabe worldview, non-human lifeforms are regarded as 'our older relatives – the ones who came before [us humans] and taught us how to live' (LaDuke, 2016, p. 18). In a similar spirit, scholar Duane Champagne (Turtle Mountain Chippewa) asserts that 'Humans are only one of many spirit beings on earth and have no exceptional role to play in the cosmic community. All beings should be respected and their roles within the cosmos comprehended' (2004, p. 312). This deep respect and the notion of being in relationship with animals are reflected not only in the stories of numerous Indigenous communities, but in Indigenous clan and kinship systems that often prescribe certain stewardship roles and complex obligations that particular clan members must maintain in relation to specific animals (or plants).

Approaching animals through these Indigenous-grounded understandings thus no longer emphasizes notions of equality, rights or justice, but instead radically shifts the frame of reference towards deep relationships, respect and obligations, with the understanding that animals have exceptional abilities that are different from those of humans; at the same time, every living being in the cosmos has its place, its role, its own Dreaming story and its reason for being, and a vitally important aspect of the human experience is in maintaining good relations with all beings. In embracing a resurgence-based paradigm in approaching the role of Indigenous interpretation in animal-based tourism experiences, new possibilities and priorities in animal-based tourism experiences can potentially be opened and this chapter takes a first step in this journey of exploration.

Methodology

Given the author's collaboration with the Indigenous leadership of the Citizen Potawatomi Nation (CPN) Aviary, her own background as a Cherokee Nation citizen of Oklahoma and her focus on Indigenous interpretations in animal tourism, this project is strongly positioned within a critical Indigenist methodological approach. Denzin *et al.* write that what defines this approach is its grounding purpose and motivation that 'begins with the concerns of Indigenous people' (2008, p. 2). Further, Indigenous scholars, including Wilson (2008) and Smith (2012), have pushed for Indigenous and decolonizing methodologies, including storywork and autoethnography, that serve as culturally grounded approaches to deeply explore knowledge, self, relationships and complex phenomena in ways that push back against and offer alternatives to Western forms of inquiry.

The author's communication with the CPN Aviary began in the latter half of 2021 and collaboration took place through much of 2022, even as certain visitor restrictions were still in place due to both COVID-19 and prevention measures against avian influenza. While the author acknowledges her own positionality as a fellow Indigenous Tribal citizen of Oklahoma, the author also recognizes that she is not a member of CPN, cannot speak on behalf of CPN perspectives and traditions, and is also a fellow learner in terms of understanding Potawatomi cultural approaches. It is toward this goal of recognizing and uplifting Potawatomi experiences and perspectives that the methodology of Indigenous storywork has deliberately been chosen.

Following on from this rootedness in Indigenous concerns and the aim to uplift the experiences of Indigenous people, this research adopts and engages with Indigenous storywork as its central methodological tool. Reflecting on the potential for learning and relationship-building that is inherent in Indigenous storywork, or the sharing of stories, scholars

Archibald (Sto:lo and St'at'imc) and Parent (Nisga'a) point to seven foundational principles underlying storywork: respect, responsibility, reverence, reciprocity, holism, inter-relatedness and synergy (2019, p. 3). Given the strong resonance between these foundational concepts in both the posthumanist and resurgence theoretical frameworks applied, as well as found within the animal-based tourism stories shared throughout the research, the importance of Indigenous storywork as integral to this chapter's methodology is evident.

Background: Oklahoma's Tribal Nations and Indigenous Connections with Eagles

The state of Oklahoma in the southern USA is home to 39 different Federally recognized Tribal Nations (Fig. 9.1). For a small handful of these Tribes, such as the Wichita and Caddo Tribes, Oklahoma has always been their ancestral homelands (Clark, 2020). Yet for the vast majority of Tribal Nations, ranging from the Citizen Potawatomi and Cherokee Nations to the Modoc, Cheyenne-Arapaho and Choctaw Nations, these Tribes were forcibly removed by the US Government to the lands designated as 'Indian Territory' – what is now the state of Oklahoma – over the course of the 19th century (Strickland, 1980; Clark, 2020). In spite of centuries of traumatic processes of forced removals, land dispossession, assimilation and assaults on culture, language and lifeways, the Tribal communities of Oklahoma have rebuilt, persevered and are in the process of revitalizing Tribal traditions. For many Tribes, this resurgent cultural re-connection has focused on language, food sovereignty and traditional community gatherings; for others, it has entailed a commitment to re-connecting with animals that hold traditional significance. While some Tribes such as the Cheyenne-Arapaho and Quapaw Nations have been raising buffalo herds and others such as the Chickasaw Nation have a Ranger programme, three Oklahoma Tribal Nations have made a concerted effort to serve as

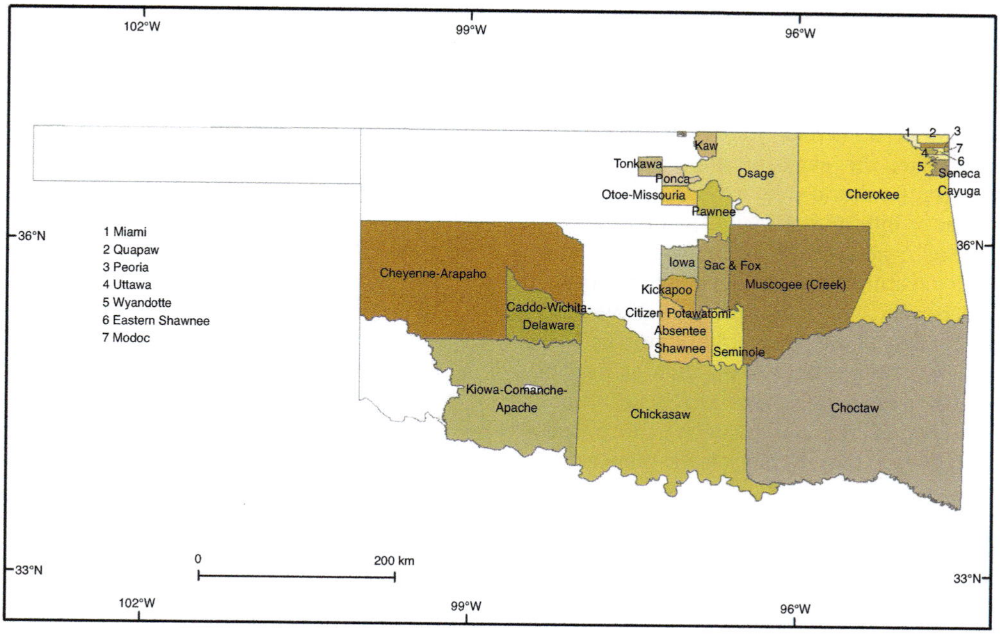

Fig. 9.1. Map of the former Tribal Reservations throughout the state of Oklahoma, many of which still represent current Tribal Jurisdictional Areas/Lands. The Citizen Potawatomi Nation's jurisdictional area is in the central-southern part of the state and in many areas overlaps with that of the Absentee Shawnee Tribe. (By Maria de Lourdes Godínez Calderón.)

stewards of eagles. These three Tribes include the Iowa Nation, Comanche Nation and the Citizen Potawatomi Nation.

The original homelands of the Citizen Potawatomi Nation (CPN), the central Tribe of focus in this chapter, covered much of the western Great Lakes region in the USA, including the present-day states of Michigan and Wisconsin. Long allied with and culturally related to the Ottawa and Ojibwe/Anishnaabe Nations of the Great Lakes area, Potawatomi people had since time immemorial depended on agriculture, hunting, berry and wild rice gathering, as well as maple tree tapping for their sustenance. Migrating further south to what is Ohio and Indiana through alliances, warfare and the fur trade, and later compelled through treaty cessions and forced removal from their northern homelands on to a reservation in Kansas, the Potawatomi would finally settle in their last and current land base in Oklahoma in 1861 (Clark, 2020, p. 319). The CPN is the only Potawatomi Tribe in Oklahoma and one of nine Potawatomi Tribes across the USA. With a population of about 30,000 members, the CPN is the largest of all the Potawatomi Tribes in the country and since the 1970s has grown its economic activities to include numerous enterprises that help to support social and cultural endeavours of the Tribe.

For numerous Tribes throughout Oklahoma, the eagle occupies an extremely important cultural role, particularly as a messenger or intermediary with the Creator. Given the eagle's ability to fly higher than nearly any other creature on earth, this characteristic is understood as a unique gift that allows eagles the honour of carrying prayers, messages and watching over earth beings, such as humans, to ensure their safety and well-being. Among members of the Iowa Nation, the belief is strongly held that the eagle is the only creature to have ever seen the face of the Creator (Iowa Tribe of Oklahoma, nd). However, the deep cultural meanings attached to the eagle are certainly not uniform across the Tribal Nations, with particular Tribes in Oklahoma maintaining a belief that only men are to come into contact with eagles and eagle feathers given that eagles are birds of prey and those energies are hostile to the fertility and life-giving nature of women (pers. comm.). For numerous other Tribes, however, these gender-based prohibitions do not exist and eagle feathers continue to play an extremely important role in ceremony, prayer, as well as adorning the regalia of Native people who are newly returned veterans, dancing in the community or even added to the cap worn by graduating Native students. For the Citizen Potawatomi Nation, eagles are understood as the creatures capable of carrying prayers from humans to the Creator, which they have always done for people and continue to do into the present-day. Moreover, eagles occupy a central role in helping the Tribe to reconnect with and reawaken numerous cultural traditions that had been dormant or severed due to the impacts of colonization and assimilation (Citizen Potawatomi Nation Cultural Heritage Center, nd).

Against this context of deep Tribal ties with eagle relatives, along with the painful history and cultural assaults faced by Native Nations across the country, eagles in the USA also sadly faced deep threats to their existence throughout the 20th century, prompting the passage of the Bald and Golden Eagle Protection Act of 1940. By the 1980s, the eagle population in the USA had declined drastically due to contamination of food sources, illegal killing and widespread habitat destruction (US Fish & Wildlife Service, nd b). Thanks to the efforts of the Protection Act, along with other efforts undertaken by the Federal Government, including the banning of DDT chemicals impacting eagle food and the Endangered Species Act of 1973, the government estimated that eagle populations had recovered and grown sufficiently to allow them to be taken off the Endangered Species List in August 2007 (US Fish & Wildlife Service, nd b).

In recognition of the deep cultural and spiritual ties that Native Nations have with eagles, the US Fish & Wildlife Service created the National Eagle Repository in the early 1970s to be able to store and distribute eagle parts, such as eagle feathers, to enrolled members of Tribal Nations. These unique rights held by members of federally recognized Tribes to be able to own and possess eagle parts are further guaranteed

Fig. 9.2. An image of the Permit received by CPN Eagle Aviary from the US Fish and Wildlife Service that allows CPN to 'receive, possess, distribute and transport eagles that are found dead within Indian Country ... for Indian religious purposes'. This Permit is housed in a display at the CPN Cultural Center. (Photo Bobbie Chew Bigby.)

under Indian Religious Use Permits; for people who are not enrolled members of Tribes, possessing eagle parts can result in jail time and costly fines. Yet in spite of the overall success of eagle population recovery across the US over the last decades, numerous eagles still face threats from human expansion into habitat areas, along with injury from wind farms and other human-made technology. In response to this, Tribal Nations and communities have stepped up to assume the role of caretaker and provide spaces for eagle rehabilitation. This movement of Tribal Eagle Aviaries has been enabled through the creation of the Tribal Eagle Aviary Permit, which authorizes Tribes 'the long-term possession of live, non-releasable eagles to be used in religious ceremonies and to collect and distribute naturally molted feathers to members of federally recognized tribes' (US Fish & Wildlife Service, nd a). The Zuni Pueblo opened the first Tribal Aviary in the USA in 1999 to serve as a sanctuary for injured eagles. The Iowa Nation was the first Tribe in Oklahoma to receive a Tribal Aviary Permit in 2005, opening their Grey Eagle Snow Eagle House the following year, which in pre-COVID times has welcomed more than 10,000 visitors annually. In Oklahoma, the Comanche Nation were the next to receive the Permit in 2009, followed by CPN in 2012 (Fig. 9.2). All of these aviaries work closely together in their efforts for eagles and all engage in educational outreach activities that have included welcoming visitors.

On Tour and in Conversation with Jennifer Randell, Leader of the CPN Tribal Eagle Aviary

This section gives an overview of the stories shared and conversations had with Jennifer Randell, the leader of the CPN Aviary tours and a Citizen Potawatomi Tribal member herself. Jennifer is the co-founder of the Aviary and also the Manager. Unless otherwise specified, the primary narrating voice in this section belongs to Jennifer and is based on the tour and conversations had with the author.

The building of CPN's Tribal Eagle Aviary programme

My family's own journey with eagles is at the heart of the story of how CPN's Eagle Aviary started. In 2007, after the untimely passing of my mother, my sister and I were gifted an eagle feather by our Chief to honour her memory, as well as the contributions that my family has made to our CPN community over the years. This gifting of the eagle feather began to lead my sister and me on a journey of connecting more deeply with eagles and other birds, as my mother had wanted to do before her passing. My sister and I headed up into Cherokee Country in Oklahoma near the Illinois River where there is a big concentration of eagle nesting sites that are great for viewing. This trip to the river was another inspirational moment nudging us more in the direction of seeing just how special these eagle relatives are, how they are facing threats to their safety and well-being, and how we as humans must step up to be stewards for eagles, just as they have watched over us.

Inspired by these meaningful moments of connection with the eagles, my sister and I gradually began learning more about some of the other Tribal-led Eagle Aviary projects going on in other parts of Oklahoma, such as with the Iowa and Comanche Nations, as well as out among the south-western Native Tribes. Given the importance of this project to our hearts, we approached our CPN Tribal Administration and Chief to ask about the possibility for our Tribe to open our own Eagle Aviary in looking after these birds. We were thrilled when we were met with support from our Chief and the rest of the Tribe. Slowly, we began to build this Aviary programme, which required everything from building relationships with the Fish & Wildlife Service, working closely with an architect and planning for the building of the Aviary and locating fresh sources of eagle food on a daily basis! We were very privileged to be granted a Tribal Eagle Aviary Permit by the US Fish & Wildlife Service in 2012, when we were finally able to open. Since this time, we have received funding from FWS; above all, it has been the support from our Tribal Nation that has been instrumental in bringing this Aviary project into being. We have also always worked very closely with our sister Tribal Aviaries in Oklahoma that belong to Iowa and Comanche Nations.

Since our Aviary opened in 2012, we have been able to help save and be the caretakers for 14 different eagles. Our Aviary's primary goals are first, to keep the eagles alive; second, to re-introduce our Tribal people to the eagle and the sacred relationship we have always had with these eagle relatives; finally, the third goal of the Aviary is to collect naturally moulted feathers from our eagles that are then distributed to Tribal Members based on request and spiritual needs. While caretaking for the eagles is the central concern, we feel that this motivation of sharing and educating people about eagles, helping to re-inspire relationships of connection between humans and these birds, is of such deep importance. This is why we put special focus on our outreach programmes and welcoming visitors to the Aviary, especially students from different schools who can come here to learn. A central priority for us is making sure that people can learn about eagles and that our Tribal members can forge their own relationships to this special relative, just as we have in our journey of building the Aviary (Fig. 9.3).

Reawakening connections to culture and eagles through the Aviary

Here at the CPN Aviary, we take in these injured eagles as family. Going back to our Potawatomi cultural beliefs, these eagles are our relatives and have long served as our prayer warriors. We believe strongly that our eagle relatives represent our physical connection to the Creator, as the eagle is the one being on earth to carry our prayers and our thoughts to the Creator. This is why eagles are so central to our spiritual beliefs and why eagle feathers are used in ceremonies,

Fig. 9.3. CPN Eagle Aviary has its main door situated at the east, signifying new beginnings, and three metal panels shaped like eagle feathers cover the Aviary at top. (Photo Bobbie Chew Bigby.)

in smudging people so that they will be cleansed and in dance. We know that the eagle has long been a protector for us as humans and thus, it is our obligation and responsibility to look after the eagles that need us. We also believe that it has been the eagle that has been the creature to help bring back many of our Potawatomi cultural traditions that had been overlooked due to the pressures of assimilation.

This deep sense of connection also influences how we go about caring for our eagles. For example, one of the males and females got together for mating and we knew that we needed to protect this couple and give them their space separate from the other eagles during nesting time. We were afraid that if we disturbed this eagle family that this action might also disturb families in our own community among the Tribe. Our connections with eagles are strong and often mirrored in our human relationships. For each eagle that comes to our Aviary, they are given a name in the Potawatomi language, allowing us to begin using our language again. But we don't give them a name once they arrive – instead we wait about four to six months until after we can observe their personality to choose the name that suits their individual character.

The design of the Aviary itself has also been part of our own cultural reawakening (see Fig. 9.3). We worked together with the architect to design a space that was top of the line in serving the needs of the eagles, but also a place that would reflect our cultural values. The Aviary is a half-circle shape that reflects the layout of our prayer circles and powwow grounds, the most important gathering places for our community where fellowship takes place. We also specifically placed the door at the east, where the sun rises and where new beginnings start from. This door allows our visitors to enter from the east when they visit the eagles. To the left and right of the doorway, we also placed two benches for people to sit and look at the eagles from an appropriate distance, but also for those Tribal members who come here to pray with the eagles. This Aviary is a space for our eagle relatives and our culture (Fig. 9.4).

Fig. 9.4. Two of the eagles in the Aviary stand next to the water. (Photo Bobbie Chew Bigby.)

Discussion

Through these stories offered by Jennifer on the tour of the CPN Aviary, it becomes apparent that visitors to this eagle sanctuary are afforded an experience that engages not only with eagles but also offers an in-depth and immersive look at Potawatomi heritage and culture. While birds are at the centre of the tour and certainly stand as the motivating draw for visitors to come to the CPN Aviary, it is clear through the thoughts and stories offered by Jennifer on the tour that all aspects of the experience are influenced and informed by cultural beliefs, stories, practices and heritage traditions. It is through this rich cultural narrative and worldview that visitors – Tribal citizens and non-Indigenous alike – are invited not only to witness connections with eagles, but to have the opportunity to (re) establish relationships with eagles and see these birds for the incredible and culturally significant gifts that they possess. Through this culturally grounded lens for experiencing and connecting with animals, this tour has an important focus on the involvement and transformation of visitors themselves, in inspiring new ways of understanding and being in relationship with animals, as well as different Indigenous cultural beliefs.

Approaching the snapshot of CPN Aviary tours through the lens of posthumanism, several elements of the tour experience appear to resonate with some of the philosophy's key tenets. Jennifer's assertion that eagles are to be highly respected and cared for as kin and as prayer warriors would certainly link strongly to the promotion of a more egalitarian approach towards animals that is advocated by posthumanist thought, actively turning away from anthropocentric positioning and towards 'extending equal treatment to humans and animals in the interaction' (Cohen, 2019, p. 422). On another level, the setup of the CPN Aviary first and foremost as a sanctuary for eagles that are injured and cannot be released back into the wild, with educational tourism activities a related but secondary priority to the task of caring for these injured birds, also aligns strongly with the posthumanist ethic that discourages the enslavement or outright commodification of these eagles that are on view for visitors. This ethic of care offered by the CPN Aviary's approach parallels posthumanist understandings of non-human animals that recognize their inherent value and thus advocates for utmost consideration of non-human animal welfare. Furthermore, the CPN Aviary tour experience can be seen to fit strongly within the posthumanist framework that prioritizes 'hybridized modes of tourism', including educational experiences that do not simply look to the Aviary as a place of entertainment or recreation for visitors (Cohen, 2019, p. 421), but instead as a site for active learning, education and stewardship training.

Yet in spite of the CPN Aviary tour's overlaps and resonances with posthumanist thought, the layers of Indigenous Potawatomi cultural interpretation offered in the tour also present some limits, if not fundamental differences, to the complete relevancy of posthumanism to this animal tourism experience. Through the Aviary tour that Jennifer leads, eagles are presented not only as animals deserving of respect and rights such as care and stewardship; instead, layers of heritage-based knowledge and cultural perspectives are shared that cast these eagles in the light through which they have traditionally always been viewed by Potawatomi people – as relatives, prayer warriors, protectors, bearers of culture and a physical connection with the Creator. This deep emphasis on the cultural and spiritual ways of viewing these birds – particularly their non-human and exceptional qualities – and being connected to them, with the encouragement among Tribal members and non-Indigenous visitors alike to find or rediscover their own connections with these eagles, would point to an understanding of this tour that puts it at the limits of posthumanism. Whereas posthumanism is centred in what is termed a 'flat ontology' where no actor or component in the system has primacy or standing above another (Law, 2004), these Potawatomi beliefs that are imparted during a tour experience do point to recognition of specific, unique and superior skills that are possessed by the eagle, and not by humans. In other words, eagles are not considered to be interchangeable with people; eagles instead occupy a distinct and highly respected role in the interconnected, living cosmos in large part because their unique gifts and abilities are different from those of humans.

Yet even though humans and eagles occupy different roles and responsibilities in the living environment, from a traditional Potawatomi perspective they are deeply connected to one another – as well as to all other aspects of creation – as literal relatives. This approach of relatedness and deep interconnection is also emphasized throughout this tour experience in the space that is created for relationship building with birds. These connections are facilitated whether it is through the sharing of Potwatomi stories of traditional eagle–human relationships or the central bench inside the CPN Aviary that allows visitors a place of prayer, reflection and communion in the presence of eagles while maintaining a safe level of distance for eagle welfare. These practices underscore the fundamental importance of relationship-building with bird relatives and that an 'ontology of connection', as articulated by Bawaka Indigenous tour leaders, as well as through theories of relationality and obligation (O'Neill, 1997; Donaldson and Kymlicka, 2011) is possible for humans, who can become exposed to this ontology and 'way of being' on a tour.

While posthumanist thought is commendable for its destabilization of hierarchies, boundaries and divisions that have traditionally pitted human against non-human within Western thought (Jackson, 2013; Cohen, 2019), it appears to approach ethical considerations for non-human animals – such as inherent value and welfare – from a stance that does not acknowledge the fundamental importance of culture and heritage in determining what ethical values and actions look like. Ethical approaches, whether promoted by posthumanism or any other philosophies, must be considered in relation to the cultural contexts and perspectives from which they arise (Grimwood *et al.*, 2018). When this cultural grounding is overlooked or divorced from ethical frameworks, important questions must be asked. First, in light of this destabilization and flattening, what exactly is left in place in regard to human and non-human relations? And second, where do philosophies that have traditionally always given primacy to relationships of stewardship and respect between humans and non-humans, such as those of Potawatomi people and numerous other Indigenous groups, fit in relation to posthumanism? In addressing these queries and the fundamentally different starting point on which Potawatomi interpretations of eagles within their Aviary tours are based, the discourse of resurgence might help to offer a more solid and culturally grounded framework for reflection and reasoning, as seen in Table 9.1.

For a visitor on the CPN Aviary tour, the starting point is not coming face-to-face with the eagles immediately. Instead, it begins with the storytelling and conversation with Jennifer who is able to share the fundamental 'starting point' of the tour – the Potawatomi cultural values and practices that are centred around deep relationships and obligations of stewardship with eagles (Fig. 9.5). These values and practices can be understood as a form of 'grounded normativity' or the place-based ethical systems of knowledge underlying Indigenous cultures (Coulthard and Alfred, 2014).

Embedded within this Potawatomi grounded normativity lies the objective of visitors (re)connecting with eagles in meaningful and respectful ways. In focusing closely on the process of (re)connection through relationships of care, respect, stewardship, reciprocity, obligation and even wonder and curiosity towards eagles, a strong link between this 'ontology of connection' and the fundamental approach of resurgence can be observed. After all, resurgence prioritizes Indigenous connections with traditional culture, lands and waters, as well as the local community that encompasses both human and non-human beings (Corntassel,

Table 9.1. A comparison of posthumanist and Indigenous Potawatomi approaches.

Posthumanist approach	Indigenous Potawatomi (CPN) approach at CPN Aviary
	Areas of resonance
Promotes egalitarianism.	Eagles are respected, cared for as kin and relatives.
Recognition of inherent value in non-human animals.	Strong cultural values associated with eagles; eagles viewed as kin and prayer warriors.
Consideration for non-human animal welfare and interests.	Welfare of eagles is promoted through direct care, stewardship and education.
Promotes hybridized modes of tourism beyond purposes of recreation and entertainment.	Caring for birds comes first before educational tours and visitor connections.
	Areas of divergence
Egalitarianism points to a 'flat ontology' where no component in living systems has primacy.	Each component of living systems has unique gifts and ways of being that are different; eagles possess skills that are different from and superior to humans on some fronts. This ontology is complex, layered and interconnected.
Egalitarianism implies a destabilization or flattening of hierarchies.	Different natures and gifts of living beings are honoured and celebrated. Through a wider ontology of connection, relationships between different beings are managed through various responsibilities, roles and obligations. Indigenous resurgence paradigm helps reawaken this ontology of connection and relationships.
Posthumanism is embedded within notions of rights, interests and egalitarianism that are largely informed by Western normativities and conceptual thinking. Other non-Western layers of meaning and conceptualization are often minimized or cultural considerations excluded.	Potawatomi cultural values focused on stewardship, responsibilities, relationships, ethics of care, traditional language and resurgence of connections stand at the core of this approach.

Fig. 9.5. Ideas of Indigenous resurgence and grounded normativity.

2021). Resurgence does this by working within Indigenous worldviews, normativities and languages themselves, as opposed to borrowing from or working towards objectives born from Western conceptualizations, such as notions of justice, rights or egalitarianism. Resurgence can thus be seen to uplift and centre the Indigenous cultural perspectives that are at the heart of the tour first and foremost, rather than overlooking these voices, or relegating them as an additional element that makes the experience unique or different from other animal encounters. This cultural uplifting is an important distinction of resurgence theory as opposed to posthumanist lenses or even relational theories that similarly prioritize human–animal relationships but leave out the rich layers of cultural meaning.

A closer examination of all the elements of the tour help to reveal the multifaceted ways in which Indigenous-led resurgence is embodied. Beyond the central activity of storytelling led by Jennifer, the careful Aviary design and practices done to care for eagles also reflect an ethos of resurgence that is grounded in cultural practice. Visitors are able to hear and use Potawatomi words as eagles are given Potawatomi names, helping to reinvigorate an Indigenous language that has come close to being lost due to the impacts of colonization and assimilation. Entering the Aviary through the main door at the east, visitors to the eagles are able to approach them from the direction of new beginnings, an important symbolic gesture in approaching these eagles to (re)establish relationships. Gathered together in the Aviary that takes the shape of the powwow grounds and prayer circles, human visitors are able to share a space with eagles in the Aviary that is reminiscent of the traditional spaces for fellowship used among Potawatomi people. Viewed through this lens of resurgence, Potawatomi cultural interpretation, perspectives and practices are put at the centre of the tour experience, allowing the opportunity for visitors to be in relationship with eagles that come to be seen not just as injured birds, but as relatives, protectors and living beings that play an extraordinary role in the web of life.

Conclusion

For visitors departing the CPN Aviary at the conclusion of a tour, Jennifer makes clear that visitors are welcomed back and encouraged to return. This encouragement to come back can be interpreted in different ways, but seen through the lens of the relationship building that is at the heart of the tour experience, it can be understood that these Aviary 'tours' really never come to an end. If the ultimate goal and outcome of most tours is the (re)establishment of deep, meaningful and respectful relationships between eagles and humans, with a deeper appreciation gained of the eagle's magnificent abilities and the ways that human–eagle interactions can be reciprocal, then the Aviary tour really only represents the beginning of this journey.

Acknowledgements

The author would like to give her deep gratitude to Jennifer Randell, Bree Dunham and all of the beautiful eagles at the CPN Eagle Aviary for being wonderful hosts. Many thanks as well to the staff and birds at the Iowa Nation's Grey Snow Eagle House, as well as Iowa Elder and Matriarch Joyce Big Soldier for sharing their time and thoughts about eagles. A big thank you is also owed to Maria de Lourdes Godínez Calderón, who helped to create the map in Fig. 9.1, based on information at: https://www.odot.org/hqdiv/p-r-div/maps/shp-files/ (accessed July 2023). The author would also like to acknowledge that this work is supported by a Social Sciences and Humanities Research Council of Canada Insight Grant (435-2018-0616) entitled, "Unsettling Tourism: Settler Stories, Indigenous Lands, and Awakening an Ethics of Reconciliation."

References

Archibald, J. and Parent, A. (2019) Hands back, hands forward for indigenous storywork as methodology. In: *Applying Indigenous Research Methods,* 1st edn. Routledge, Abingdon, UK, pp. 3–20. DOI: 10.4324/9781315169811.

Bekoff, M. and Sherman, P.W. (2004) Reflections on animal selves. *Trends in Ecology & Evolution* 19(4), 176–180. DOI: 10.1016/j.tree.2003.12.010.

Champagne, D. (2004) Tribal capitalism and native capitalists: multiple pathways of native economy. In: Hosmer, B. and O'Neill, C. (eds) *Native Pathways: American Indian Culture and Economic Development in the Twentieth Century*. University Press of Colorado, Boulder, Colorado.

Citizen Potawatomi Nation Cultural Heritage Center (n.d.) 'Aviary' Citizen Potawatomi Nation Cultural Heritage Center (website). Available at: https://www.potawatomiheritage.com/aviary/ (accessed 15 April 2022).

Clark, B. (2020) *Indian Tribes of Oklahoma: A Guide*, Second Edition. University of Oklahoma Press, Norman, Oklahoma.

Cohen, E. (2019) Posthumanism and tourism. *Tourism Review* 74(3), 416–427. DOI: 10.1108/TR-06-2018-0089.

Corntassel, J. (2021) Life beyond the state: regenerating indigenous international relations and everyday challenges to settler colonialism. *Anarchist Developments in Cultural Studies* 1, 71–98.

Coulthard, G.S. and Alfred, T. (2014) *Red Skin, White Masks: Rejecting the Colonial Politics of Recognition*. University of Minnesota Press, Minneapolis, Minnesota. DOI: 10.5749/minnesota/9780816679645.001.0001.

Denzin, N.K., Lincoln, Y.S. and Smith, L.T. (2008) *Handbook of Critical and Indigenous Methodologies*. Sage, Los Angeles, California. DOI: 10.4135/9781483385686.

Donaldson, S. and Kymlicka, W. (2011) *Zoopolis: A political theory of animal rights*. Oxford University Press, Oxford, UK.

Edelman, D.B. and Seth, A.K. (2009) Animal consciousness: a synthetic approach. *Trends in Neurosciences* 32(9), 476–484. DOI: 10.1016/j.tins.2009.05.008.

Grimwood, B., Caton, K. and Cooke, L. (eds) (2018) *New Moral Natures in Tourism*. Taylor & Francis, London. DOI: 10.4324/9781315265117.

Guia, J. (2021) Conceptualizing justice tourism and the promise of posthumanism. *Journal of Sustainable Tourism* 29(2–3), 503–520. DOI: 10.1080/09669582.2020.1771347.

Iowa Tribe of Oklahoma (n.d.) 'Grey snow eagle house: Iowa tribe of Oklahoma'. Iowa Tribe of Oklahoma (website). Available at: http://eagles.iowanation.org/

Irvine, L. (2007) The question of animal selves: implications for sociological knowledge and practice. *Qualitative Sociology Review* 31(1), 5–22. DOI: 10.18778/1733-8077.3.1.02.

Jackson, Z.I. (2013) Animal: new directions in the theorization of race and posthumanism. *Feminist Studies* 39(3), 669–685. DOI: 10.1353/fem.2013.0024.

Kline, C. (2022) Socialising animal-based tourism. In: Higgins-Desbiolles, F., Doering, A. and Bigby, B.C. (eds) *Socialising Tourism: Rethinking Tourism for Social and Ecological Justice*. Routledge, London, pp. 195–213. DOI: 10.4324/9781003164616.

LaDuke, W. (2016) *All Our Relations: Native Struggles for Land and Life*. Haymarket Books, Chicago, Illinois.

Law, J. (2004) And if the global were small and incoherent? Method, complexity, and the Baroque. *Environment and Planning D: Society and Space* 22(1), 13–26. DOI: 10.1068/d316t.

O'Neill, O. (1997) Environmental values, anthropocentrism and speciesism. *Environmental Values* 6(2), 127–142. DOI: 10.3197/096327197776679121.

Ruhanen, L. and Whitford, M. (2019) Cultural heritage and indigenous tourism. *Journal of Heritage Tourism* 14(3), 179–191. DOI: 10.1080/1743873X.2019.1581788.

Smith, L.T. (2012) *Decolonizing Methodologies: Research and Indigenous Peoples*, Second edition. Zed Books, London.

Starblanket, G. *et al*. (2018) Resurgence as relationality. In: Corntassel, J. (ed.) *Everyday Acts of Resurgence: People, Places, Practices*. Daykeeper Press, Olympia, Washington, pp. 28–32.

Strickland, R. (1980) *The Indians in Oklahoma*. University of Oklahoma Press, Norman, Oklahoma.

US Fish & Wildlife Service (nd a) 'Bald and Golden Eagle Management'. US Fish & Wildlife Service (website). Available at: https://www.fws.gov/library/collections/bald-and-golden-eagle-management (accessed 15 April 2022).

US Fish & Wildlife Service (nd b) '3-200-78: Native American Tribal Eagle Aviary'. US Fish & Wildlife Service (website). Available at: https://www.fws.gov/service/3-200-78-native-american-tribal-eagle-aviary (accessed 15 April 2022).

Walker, K. and Moscardo, G. (2016) Moving beyond sense of place to care of place: the role of Indigenous values and interpretation in promoting transformative change in tourists' place images and personal values. *Journal of Sustainable Tourism and Indigenous Peoples* 24(8–9), 1243–1261. DOI: 10.1080/09669582.2016.1177064.

Welsch, W. (2017) 'Postmodernism—posthumanism—evolutionary anthropology'. *Journal of Posthuman Studies* 1(1), 75–86. DOI: 10.5325/jpoststud.1.1.0075.

Wilson, S. (2008) *Research is Ceremony: Indigenous Research Methods*. Fernwood Pub, Black Point, Nova Scotia, Canada.

Wright, S., Suchet-Pearson, S., Lloyd, K., Burarrwanga, L.L. and Burarrwanga, D. (2009) 'That means the fish are fat': sharing experiences of animals through Indigenous-owned tourism. *Current Issues in Tourism* 12(5–6), 505–527. DOI: 10.1080/13683500903042907.

Zeppel, H. (2006) *Indigenous Ecotourism Sustainable Development and Management*. CABI, Wallingford, UK. DOI: 10.1079/9781845931247.0000.

Zeppel, H. (2010) Managing cultural values in sustainable tourism: conflicts in protected areas. *Tourism and Hospitality Research* 10(2), 93–115. DOI: 10.1057/thr.2009.28.

Zeppel, H. and Muloin, S. (2008a) Aboriginal interpretation in Australian wildlife tourism. *Journal of Ecotourism* 7(2 & 3), 111–131. DOI: 10.2167/joe0226.0.

Zeppel, H. and Muloin, S. (2008b) Conservation benefits of interpretation on marine wildlife tours. *Human Dimensions of Wildlife* 13(4), 280–294. DOI: 10.1080/10871200802187105.

Zeppel, H., Muloin, S. and Higginbottom, K. (2003) Kangaroo or gangurru? Indigenous wildlife interpretation in Australia. In: Black, R.S. and Weiler, B. (eds) *Interpreting the Land down under: Australian Heritage Interpretation and Tour Guiding*. Fulcrum Publishing, Golden, Colorado, pp. 165–185.

10 A Posthumanist View on Theming the Axolotl for a *Pueblo Mágico*

Gino Jafet Quintero Venegas[1,*], Álvaro López López[1] and Carol Kline[2]
[1]Universidad Nacional Autónoma de México, Mexico City, Mexico; [2]Appalachian State University, Boone, North Carolina, USA

Abstract
Known for the production of handcrafted Christmas-tree glass baubles and a festival that celebrates life, Chignahuapan, a small town in central Mexico, was designated as '*Pueblo Mágico*' ('Magical Town') in 2012. As with any town that has the distinction, promotional images of the town have been disseminated nationwide to attract scores of visitors. To compete with other *Pueblos Mágicos*, Chignahuapan has chosen to adopt the axolotl as a trademark of its tourist offer and local cultural heritage. In an effort to reinforce the axolotl's presence as an element of local identity and belonging, images of the amphibian have been distributed throughout the town, while live specimens are displayed in several spaces. The axolotl is associated with the idea of regeneration, which has influenced the way both local inhabitants and tourists conceive of the town. As such, Chignahuapan has the potential to become a zoopolis, however additional work is needed to dismantle innate hierarchical structures.

Introduction

Chignahuapan ('Over the nine waters' in Nahuatl) is a small town in the north-eastern part of the state of Puebla. In 2012, it got its title as *Pueblo Mágico* in recognition of its thermal springs resort –famous for the healing properties of its waters – and the production of handcrafted Christmas-tree baubles (Quintero, 2021). As with many other cases, this federal government distinction has translated into a higher flow of tourists to the town, who are intent on 'consuming' the spatial images as they see them in tourism advertisements (Gottdiener, 2000).

Generally, when a town receives the title of *Pueblo Mágico*, its historic centre undergoes a transformation so as to make it attractive to tourists. Theming is a recurring practice, and the model takes its cue from the ideal rural Mexican village so prevalent in the imaginary of most tourists. The result is a homogenization of urban landscapes, therefore through the ensuing development, each town seeks to project distinctive images of their own identity.

In Chignahuapan, the axolotl has been adopted as a symbol of the town and promoted as an element of its cultural heritage. Live specimens of the amphibian have been introduced into the local economic dynamics, and the urban landscape has been endowed with its image. Festivals in honour of the axolotl are organized regularly, which reinforces the reification – the consideration of a conscious and free living being as if it were an object or thing that is not conscious or free – of the species (Quintero and López, 2018). The exhibition and commercialization of the axolotl has been seen as an economic alternative by

*jafet@sociales.unam.mx

 Tourism, Heritage and Commodification of Non-human Animals (eds A. López López *et al.*)
DOI: 10.1079/9781800623309.0010

a number of agriculturists who, in view of the process of depeasantization going on in the region and the entire nation (Quintero, 2021), seek to find a more stable and lucrative economic activity.

This chapter analyses the reification of the axolotl in Chignahuapan as a trademark and a product for tourism consumption. Qualitative research was conducted following the triangulation strategy (Decrop, 1999), combining three techniques: (i) documentary searches and analysis of texts (Ruiz, 2012) pertaining to the spatial and temporal configuration of tourism in Chignahuapan as a *Pueblo Mágico* (Tedlock, 1991; Quintero, 2021); (ii) participant observation during visits to the town in 2019 and 2021; and (iii) discourse analysis based on the comments by users of the TripAdvisor website (Lei and Law, 2015).

Though humans share with animals a number of things in different areas of life, in the realm of anthropocentrism inter-specific relations pass usually unnoticed. Animals are rendered invisible unless they are of use to humans, for acknowledging their intrinsic interests would make humans accountable on several points (Horta, 2010). A posthuman paradigm that includes the rights of animals seeks to change this (Wolfe, 2010), i.e. to substitute the established instrumentalist mentality for a new vision of respect for the rights and interests of other animal species (Horta, 2010; Kline, 2021). Thus it is the purpose of this research to illuminate the symbolic role that animals play in tourist destinations. While there is potential for towns and regions to value their token-animals by enacting policies that protect them and programmes to educate about their role in the local ecosystem, most destinations are far from creating a space where humans and non-humans can co-exist.

The *Pueblos Mágicos* Programme and the Construction of a Tourism Imaginary

Chignahuapan has a population of 66,500 inhabitants (Data México, 2021; Fig. 10.1). Though the town already had an incipient tourism activity, its inclusion in the federal programme of *Pueblos Mágicos* in 2012 turned it into a cultural-tourism destination on account of the production of handcrafted blown-glass baubles, the Light and Life Festival, celebrated every year on the Day of the Dead (1 and 2 November), and the presence of a thermal springs spa, which attracts spa and health tourists (Quintero and López, 2018; Quintero, 2021).

The objective of the *Pueblos Mágicos* programme is to promote towns as cultural-tourism destinations based on their heritage (Rojo and Llanes, 2009; Méndez and Rodríguez, 2013). The town then undergoes a process of urban planning and conservation. In this way, it is assumed, cultural heritage will be preserved, new urban practices introduced, urban space reconfigured and reappraised, and an increase in tourism expenditures will be gained (Rojo and Llanes, 2009; Quintero, 2021).

Though the role of the local community in preserving their cultural and historic richness is emphasized (Rojo and Llanes, 2009), the process is not exempt from social inequality, spatial disparities and Disneyfication. As Velázquez (2012) maintains, the inclusion of small communities into the national and international tourist circuits leads to the commodification of public space for the benefit of visitors.

The designation of Chignahuapan as *Pueblo Mágico* in 2012 was severely criticized, if for no other reason for the fact that it was perceived as unwanted competition for neighbouring Zacatlan, a town only 20 km away (Quintero and Castro, 2022) that had received the distinction 1 year earlier. The fear that Chignahuapan could 'steal' potential overnighters from Zacatlan was a point of contention. However, time has confirmed that the two communities have been able to complement each other and benefit from the increase in tourism (Ríos, 2018). Another point of contention has been the lack of benefits realized for the majority of the inhabitants of Chignahuapan. While downtown houses have been embellished (Castro and Quintero, 2022), investments in infrastructure remain limited, notably access routes to the town and improvement of schools. The general perception by residents is that tourism has failed to improve the quality of life of the population at large (Barrios, 2016).

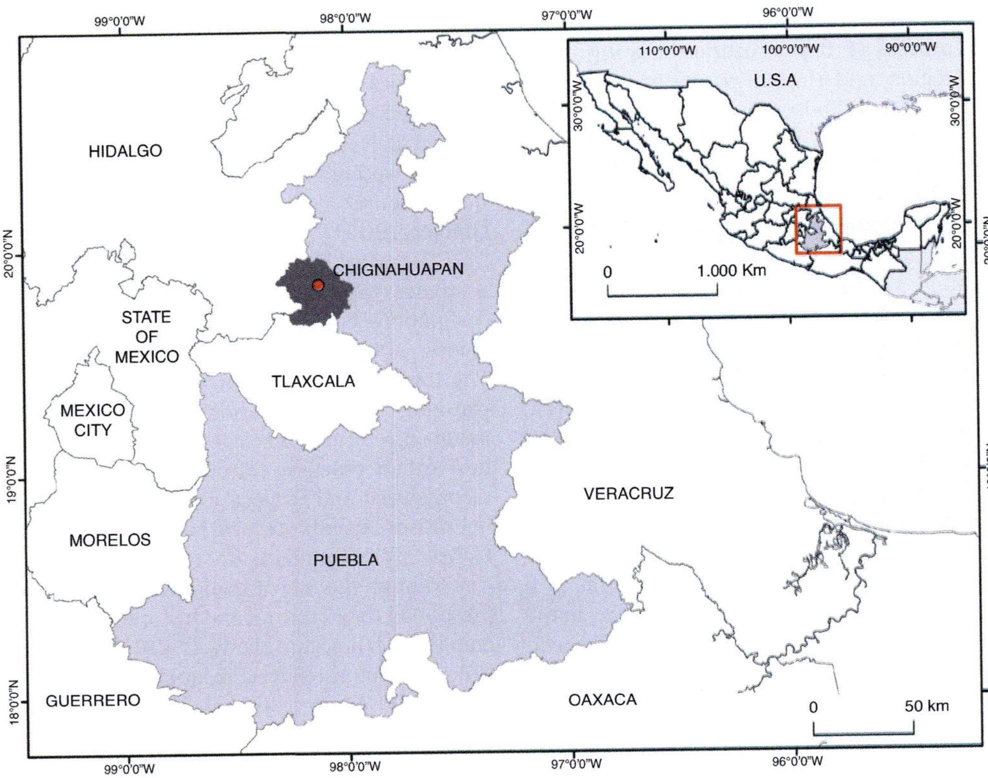

Fig. 10.1. Location of Chignahuapan. (Drawn by María de Lourdes Godínez Calderón, 2022.)

Theming Chignahuapan

Theming is the process of creating an imagined reality by stressing the qualities of a given contemporary society, using spatial symbols that can be easily recognized everywhere (Salas and López, 2019; Åstrøm, 2020). It is a creative exercise that blends imagination, technical capabilities, artistic concepts and visitors' opinions into a dynamic in which aesthetic variables are fundamental. Theming seeks to create an environment of entertainment that can be fictitious or real or a blend, offering a unique atmosphere that elicits emotional responses in consumers.

As a strategy that defines and commercializes space for consumption by tourists (Mitrasinovic, 2016), theming can also reinforce the improvement and preservation of heritage (Jouault *et al.*, 2018). Theming implies the reconfiguration of those environmental attributes that can be stimulating to the senses and incite consumption. In order to turn it into an aesthetically agreeable landscape, public space is invaded and intervened (Delgado, 2007). This social-construction process turns living spaces into merchandise, as the environment is redefined as a space for consumption. In so doing, theming makes use of spatial transformation strategies, as well as publicity and marketing (Salas and López, 2019).

The tourism trademark '*Pueblos Mágicos*' is a typical theming exercise which has brought competition among tourist destinations that attempt to distinguish themselves from one another. In the case of Chignahuapan, the unique products offered (the axolotl, the thermal springs, the baubles) allow the destination to position itself in the market with an image that is easily identifiable and hard to substitute (Ávila and Barrado, 2005). The spatial representation strategy implemented expresses the unique geographical identity of the destination, and imaginaries are invented and reinvented to create themes for tourism (Guerrero *et al.*, 2018); this theming must include elements that make the space unique, so as to ensure competitiveness, distinction and profitability (Córdoba y Ordónez, 2009). For example, Chignahuapan competes with Tlalpujahua in the selling of the Christmas baubles, and with many other *Pueblos Mágicos* with thermal springs. So, in addition to these two features, the axolotl has also become a trademark.

Underlying the implementation of the *Pueblos Mágicos* programme is the idea of representing 'Mexicanness' (Méndez and Rodríguez, 2013). The image of the axolotl is ideal to this end, for it is an animal endemic to central Mexico and an important historic and cultural referent. The Aztecs thought of the axolotl as a divine animal representing the god Xolotl (Holguin, 2019). The cultural value of the axolotl in Chignahuapan is well established, and is derived from the history of the community and the indigenous view of the world. According to the Nahua legend, the axolotl is an embodied manifestation of the god Xolotl – twin brother of the god Quetzalcoatl – who when born had a monstrous appearance and hence was doomed to be killed. He escaped this fate by plunging into the water and transformed into the well-known amphibian (De la Garza, 2013). For Chignahuapans, due to its characteristics as adaptable and being 'a survivor', Xolotl is related to the idea of movement and life, qualities derived from this traditional legend (De la Garza, 2013).

Additionally, a local family, the Familia Carbajal, started the House of the Axolotl in 2013 (now MUMAX) to preserve the axolotl, generate tourism in the city and reinforce its image as an identity symbol (Galeana, 2020). Yanin Carbajal, the daughter of the family, studied biology at National Autonomous University of Mexico. She was interviewed in 2017 and said:

> We started 4 years ago, we decided to exhibit them because we realized that most of the country do not know the characteristics of the axolotls nor have they seen them in nature. Many people want to reproduce trout and carp in this region without knowing that little by little they are destroying the axolotl. Despite the fact that some species are herbivorous, when eating plants they do not distinguish and take eggs (of the axolotl) as well. For now it is unknown how many remain in the area, yes, here there are still many places where they live naturally. In Chignahuapan the species can be preserved but we must not trust ourselves, we must continue working to avoid its extinction.
>
> (Galeana, 2017)

Thus, it was decided to use the axolotl as an identity symbol and a mechanism of community cohesion, which could also be attractive to visitors.

Conservation of the Axolotl

The axolotl (*Ambystoma mexicanum*) is a salamander of a little more than 30cm in length, which keeps certain traits of its larval stage during adulthood and is able to reproduce itself before maturity (Tompkins, 1978; Fig. 10.2). It is an unusual amphibian that fails to undergo metamorphosis and develop lungs, and keeps to its aquatic environment instead of transiting to terrestrial life (Tompkins, 1978). Though it is endemic to the lacustrine environment of Xochimilco and Chalco-Tlahuac in south-eastern Mexico City, there are two exceptional species in the state of Puebla (Flores, 2003; Gresens, 2004).

The species has been on the protected species list since 1994, and since 2010 in the 'threatened' category (Bride *et al.*, 2008; IUCN, 2010). In the last few years, its habitat has been dramatically reduced, due to several factors: (i) the expansion of Mexico City's Metropolitan Area; (ii) the draining of the lakes where the amphibian thrives in order to open up more land for construction; (iii) the domestic and industrial pollution of the lakes; (iv) the introduction of fish species for human consumption – such as carp, mojarra and tilapia – which prey on the eggs and larvae of the amphibian; and (v) the legal and illegal trade of axolotl specimens, believed to have healing properties, thanks to their ability to regenerate their own tissue (Stephan and Ensástigue, 2001; Quintero and López, 2018).

Two places in Chignahuapan have been devoted to the conservation of the axolotl: an Environmental Management Unit (UMA) in a dam located in a forested area of the municipality, and MUMAX, where individuals of this

Fig. 10.2. Axolotl (*Ambystoma mexicanum*). (Photo by G.J. Quintero Venegas.)

species are exhibited in aquariums. Souvenirs representing the axolotl – felt toys, key rings, postcards – can be bought at the museum's shop, and the money collected from the entrance fee goes to a fund to finance conservation plans.

The question remains, however, whether these conservation efforts are motivated by the charismatic qualities of the axolotl, or whether they respond to medical, ecological and bio-cultural interests (Vázquez, 2011), or perhaps some combination. Axolotl specimens are kept in laboratories to carry out research on their exceptional ability to regenerate any type of tissue in their bodies, including heart and neurons (Gardiner *et al.*, 1995). On the other hand, in the case of Xochimilco, the axolotl is considered a necessary predator endemic to the lake, which regulates the functioning of the trophic chain, and is believed to help combat anthropogenic pollution (Mendoza, 2015). Finally, axolotls are not only a key ingredient in culinary recipes for certain kinds of broths, soups, stews and tamales, but their ingestion is also believed to cure tuberculosis, asthma and bronchitis (Campos, 2016).

Chignahuapan: A Themed *Zoopolis*?

The term '*zoopolis*' has a political and a geographical meaning; it expresses the need to protect non-human animals, so that they can enjoy a good life as sentient beings in their home environments (Donaldson and Kymlicka, 2011). Donaldson and Kymlicka maintain that non-human animals can be classified, according to their habitat, as 'citizens' (domesticated animals), 'sovereign' (wild animals) or 'city dwellers' (liminal animals). Each category has corresponding specific rights and duties, and non-human animal communities must enjoy recognition as sovereign communities in the same way as human communities (Donaldson and Kymlicka, 2011; Horta, 2013).

The term is a neologism formed by two Greek words: ζώο (*zo̲o*) 'animal' and πόλις (*pólis*) 'city', and it is conceived of as a town in which human and non-human animals coexist under mutual-benefit relations. The zoopolis is an alternative version of the natural space conceived as complementary to the metropolis, where humans and non-humans could coexist (Urbanik, 2012). It can also be an urbanized space where parallel worlds (those of humans and non-humans) merge and coexist (Urbanik, 2012).

Several sectors of Chignahuapan's historic centre, where tourist consumption concentrates, are slowly turning into a zoopolis of sorts in which *Homo sapiens* individuals willingly coexist with individuals of *Ambystoma mexicanum*. Present everywhere as an element of the town's cultural heritage, the axolotl has become a synanthrope, that is, an animal species adapted to an anthropic ecosystem (Cursach *et al.*, 2012). This spatial dichotomy is characteristic of several other urban tourist spaces where humans coexist with other animal species, such as pigeons and squirrels (López and Quintero, 2021), and public policy must be developed or even considered to address the rights of these animal species.

The image of the axolotl as a symbol of regeneration has also been exploited for political purposes. In Chignahuapan, the Morena (National Regeneration Movement) Party, which enjoys a congressional majority under the government of President Andrés Manuel López Obrador, has adopted the image of an animal capable of regenerating its body to invite voters to join efforts in addressing the problems left behind by former governments (Fig. 10.3).

The urban space of Chignahuapan has been themed with the image of the axolotl. At the town's centre, for example, the visitor can see flowerpots, mosaics and other decorations displaying its image, which reinforces the species as part of the town's heritage. The town's main

Fig. 10.3. 'Regeneration without fear'. Analogy between the axolotl and local politics. (Photo by G.J. Quintero Venegas.)

tour operator uses the name of the axolotl to sell the 'Ajolotour', a guided journey of a little more than an hour in duration through the main sites of cultural importance in the town. The construction of this tourist trademark has been successful in reinforcing local identity and heritage, and has led to the potential of Chignahuapan to become a zoopolis.

Another promotional strategy common to many tourist nodes is the erection of 'brand monumental letters' with the town's name in a central location, in order to produce visual content to be disseminated in social networks and other virtual platforms (Smith, 2018). In Chignahuapan, the letters are decorated with images of the axolotl and other identity elements. Each time tourists take a photo of this monument, they contribute to reinforce the axolotl as an identity element of the town.

In addition to this, a series of murals depicting the axolotl are found throughout the town's centre. The animal is portrayed as 'protector' of the community as much in life as in death (Fig. 10.4).

Among other forms of urban theming with the axolotl as protagonist is the Ajolotiada, an annual festival celebrated to acknowledge the importance of the animal as cultural heritage and create awareness of the need to preserve it (Puebla Noticias, 2017). Apart from tourist trips on the Axolot-bus (Fig. 10.5) and visits to the Axolote Museum, visitors can attend lectures on the species' characteristics, theatre performances featuring the god Xolotl, concerts by local music groups and parades of oversize cardboard puppets (Quintero and López, 2018).

Contested Space and the Teleological Metamorphosis of the Axolotl

Paradoxically, federal conservation policies restrict the use of the axolotl for tourism purposes. The selling of specimens for consumption

Fig. 10.4. Mural representing the axolotl as protector of Chignahuapan. (Photo by G.J. Quintero Venegas.)

Fig. 10.5. The Axolot-bus. (Photo by G.J. Quintero Venegas.)

as food or medicine in the central market and other public spaces has been strictly prohibited, and its commercialization can only be effected for conservation purposes (Parra *et al.*, 2014). When a government institution regulates and restricts a traditional economic activity in favour of tourism, a competition for space is generated. As Brouder and Ioannides maintain:

> Tourism development never occurs in a vacuum and urban tourism is one issue where the complexity of the tourism economy and its place within broader regional development strategies is most obvious. The co-evolution of the various tourism paths in urban areas is especially complex and so too is the interaction of the various tourism paths with other regional economic developments. For instance, urban tourism enters the fray in matters of contested urban spaces with issues of local governance, such as privatisation of public space, moving increasingly to the fore.
>
> (Brouder and Ioannides, 2014, p. 1)

It can be said that, in Chignahuapan, the competition for space by different types of tourism has been beneficial to the axolotl. Though the amphibian is still regarded as a commodity by many, awareness of its intrinsic value as a protected species that coexists in a complex life environment is growing among the local population. A zoopolis of sorts has thus been developed (Katcher and Wilkins, 1993; Urbanik, 2012). However, due to the anthropocentric lens through which we humans perceive our relations with the natural world, this kind of arrangement passes mostly unnoticed.

There are two common ways of viewing animal species: by their intrinsic value as sentient beings (Singer, 2009; Horta, 2010), and by their extrinsic value as instruments to human well-being. The first is an act of celebration of life that acknowledges the right of animals to live according to their needs and preferences. The second is an anthropocentric view that focuses on the utility of animals to humans and imbues a utilitarian teleological finality on all other species on these terms (Madrid, 2004). Animals cease to be regarded as individuals with rights and become property (Ryder, 1989). In Chignahuapan, the axolotl is increasingly and slowly being regarded in its own right, but there is still a long way to go before this animal is no longer considered some kind of resource.

The arrival of tourism in Chignahuapan has transformed the teleological attribute of the axolotl. Before it was adopted as heritage, this animal was conceived of as a being of value for its healing and nutritional properties. Now it is regarded as an individual being whose will and basic interests cannot be ignored or violated and must instead be honoured for its role as an identity element and tourism trademark.

The MUMAX showcases the axolotl in its natural habitat, as well as providing information on the salamander's biology, ecology and cultural significance in the region. The museum also has a breeding programme for axolotls and works to promote their conservation and restoration in the wild. Visitors can see the axolotls up close and learn about the important role they play in the local ecosystem and the broader scientific community.

The role of the MUMAX in changing the perception of the amphibian cannot be overemphasized. Starting out with the objective of rescuing the species, today it is a space for display of specimens so that visitors can get acquainted with them, develop affective links and empathy towards them, and become aware of their intrinsic value as sentient beings. As Yanin Carbajal, one of the founders of the museum, said in an interview:

> We decided to exhibit them because we realized that most people have no knowledge of the characteristics of the axolotl and have never seen a specimen in their natural environment ... It has been a real challenge to be able to display them but we have made progress and we want people to get to know them and appreciate them for what they are.
>
> (González, 2020)

The MUMAX is registered at the Environment and Natural Resources Secretariat (Semarnat) as an Environmental Management Unit (UMA) for the conservation and reproduction of the axolotl. It has a games room and various spaces for the dissemination of scientific knowledge in which workshops, courses and colloquiums are offered to visitors with the aim of highlighting the value of the axolotl and raising funds for its conservation. However, exhibition and living spaces remain rather small and the funds raised from tourism are far from sufficient to improve them.

Conclusions

The case of Chignahuapan and its adoption of the axolotl as a symbol and an element of heritage is symptomatic of the economic tertiarization process in cities endowed with the *Pueblo Mágico* title. The tertiary sector of the economy plays a prominent role in theming cities and towns according to cultural narratives that are foreign to the local inhabitants; theming goes hand-in-hand with specialization. In the case of Chignahuapan, a pre-existing narrative (i.e. the legend of the axolotl) has served the purpose. But theming and the need to attract tourists have led to a competition for space between the tourism industry and the local population.

Chignahuapan has become the town of the axolotl. Countless images of the amphibian adorn the entire urban landscape: drawings, engravings, sculptures and photographs are found everywhere. These promotional urban images are attractive to tourists and, at the same time, constitute a referent to local culture. Both locals and tourists now see the axolotl as an element of the town's identity.

If it is true that tourism has helped to spur the conservation of the axolotl – a species now endangered in its natural habitat in an area too close to Mexico City to effectively be protected – such conservation is mired in a contradiction. In the MUMAX, specimens are exhibited in very small fish tanks, which not only are far from resembling their natural habitat but also drastically limit their mobility. The ethical problem of sacrificing the well-being of a sentient being in order to extract economic benefit is evident. To what extent can such a reification of a living being be justified on the grounds of having been hailed as heritage?

Though the ubiquitous images of the axolotl have helped raise awareness of the importance of the species and its conservation, they have also exploited the amphibian as an object for tourist enjoyment. In this respect, the creation of conservation institutions and the design of awareness campaigns for tourists are a desirable way to go, so that the axolotl can be appraised for its intrinsic value and freed from the humanist, hierarchical relations to which it is now subjected in tourist spaces.

Acknowledgements

This chapter is part of a research project entitled 'Depeasantization, economic tertiarization and new ruralities as a result of the insertion of rural and cultural tourism in the *Pueblos Mágicos* of Central Mexico', carried out at the National University's Social Research Institute (IIS-UNAM).

References

Åstrøm, J. (2020) Why theming? Identifying the purposes of theming in tourism. *Journal of Quality Assurance in Hospitality & Tourism* 21(3), 245–266. DOI: 10.1080/1528008X.2019.1658147.

Ávila, R. and Barrado, D. (2005) Nuevas tendencias en el desarrollo de los destinos turísticos: marcos conceptuales y operativos para su planificación y gestión. *Cuadernos de Turismo* 15, 27–44.

Barrios, E. (2016) Inseguridad pone en riesgo título de pueblo mágico de Chignahuapan. El Sol de Puebla. Available at: https://www.elsoldepuebla.com.mx/estado/inseguridad-pone-en-riesgo-titulo-pueblo-magico-chignahuapan (accessed 10 April 2023).

Bride, I.G., Griffiths, R.A., Meléndez-Herrada, A. and McKay, J.E. (2008) Flying an amphibian flagship: conservation of the axolotl *Ambystoma mexicanum* through nature tourism at Lake Xochimilco, Mexico. *International Zoo Yearbook* 42(1), 116–124. DOI: 10.1111/j.1748-1090.2008.00044.x.

Brouder, P. and Ioannides, D. (2014) Urban tourism and evolutionary economic geography: complexity and co-evolution in contested spaces. *Urban Forum* 25(4), 419–430. DOI: 10.1007/s12132-014-9239-z.

Campos, T. (2016) Axolotl: así es el mítico, y desconocido, anfibio exclusivo del valle de México. *Xataka México*. Available at: https://www.xataka.com.mx/ciencia/axolotl-asi-es-el-mitico-y-desconocido-anfibio-exclusivo-del-valle-de-mexico (accessed 10 April 2023).

Castro, A. and Quintero, G. (2022) Interacción espacial de la dinámica turística de los pueblos mágicos de Zacatlán y Chignahuapan, Puebla. *Investigaciones Geográficas* 64, 66–81. DOI: 10.5354/0719-5370.2022.67994.

Córdoba y Ordónez, J. (2009) Turismo, desarrollo y disneyzación: una cuestión de recursos o de ingenio. *Investigaciones Geográficas* (70), 33–54.

Cursach, J., Rau, J., Ther, F., Vilugrón, J. and Tobar, C. (2012) Sinantropía y conservación marina: el caso del chungungo *Lontra felina* en el sur de Chile. *Revista de Biología Marina y Oceanografía* 47(3), 593–597. DOI: 10.4067/S0718-19572012000300022.

Data México (2021) Chignahuapan. Available at: https://datamexico.org/es/profile/geo/chignahuapan#:~:text=La%20poblaci%C3%B3n%20total%20de%20Chignahuapan,%25%20mujeres%20y%20 48.4%25%20hombres (accessed 10 April 2023).

Decrop, A. (1999) Triangulation in qualitative tourism research. *Tourism Management* 20(1), 157–161. DOI: 10.1016/S0261-5177(98)00102-2.

De la Garza, M. (2013) El carácter sagrado del xoloitzcuintli entre los nahuas y los mayas. *Arqueología Mexicana* 31(125), 58–63.

Delgado, M. (2007) Ciudades sin ciudad. La tematización 'cultural' de los centros urbanos. *Antropología y Turismo. Claves Culturales y Disciplinares* 91–109.

Donaldson, S. and Kymlicka, W. (2011) *Zoopolis: a Political Theory of Animal Rights*. Oxford University Press, Oxford, UK.
Flores, J. (2003) *Axólotl: El Ajolote*. Ediciones Era, Mexico City.
Galeana, M. (2017) Los axolotes también viven en Puebla. Manati. Available at: https://manati.mx/2017/08/27/axolotes-chignahuapan-puebla/ (accessed 10 April 2023).
Galeana, M. (2020) Seis años del primer museo Mexicano del Axolote. Manati. Available at: https://manati.mx/2020/10/12/museo-mexicano-del-axolote-chignahuapan/ (accessed 10 April 2023).
Gardiner, D.M., Blumberg, B., Komine, Y. and Bryant, S.V. (1995) Regulation of HoxA expression in developing and regenerating axolotl limbs. *Development* 121(6), 1731–1741. DOI: 10.1242/dev.121.6.1731.
González, R. (2020) Promueven primer museo Mexicano del axolote en Chignahuapan. Milenio. Available at: https://www.milenio.com/politica/comunidad/promueven-al-museo-mexicano-del-axolote-en-chignahuapan (accessed 10 August 2022).
Gottdiener, M. (2000) The consumption of space and the spaces of consumption. In: *New Forms of Consumption: Consumers, Culture, and Commodification*. Rowman & Littlefield Publishers, pp. 265–284.
Gresens, J. (2004) An introduction to the Mexican axolotl (*Ambystoma mexicanum*). *Lab Animal* 33(9), 41–47. DOI: 10.1038/laban1004-41.
Guerrero, R., Pérez, M. and Alvarado-Sizzo, I. (2018) Qué´opinan nuestros turistas. Un análisis temático en TripAdvisor sobre atractivos turísticos de la ciudad de Guanajuato. *Dimensiones Turísticas* 2(3), 91–110. DOI: 10.47557/VTGU4902.
Holguin, F. (2019) La mexicanidad de Xólotl/Axolotl: maravilla de la literatura y la gastronomía. *Mitologías Hoy* 19, 147–159. DOI: 10.5565/rev/mitologias.622.
Horta, O. (2010) Igualitarismo, igualación a la baja, antropocentrismo y valor de la vida. *Revista de Filosofía* 35(1), 133.
Horta, O. (2013) Zoopolis, interventions and the state of nature. *Law, Ethics and Philosophy* 1, 113–125.
IUCN (2010) *Ambystoma mexicanum*. The International Union for conservation of nature red list of threatened species. IUCN. Available at: http://www.iucnredlist.org/details/1095/0 (accessed 10 August 2023).
Jouault, S., Enseñat-Soberanis, F. and Balladares-Soberano, C. (2018) La milpa maya en Yucatán: una transición entre la patrimonialización y la turistificación? *Revista Gremium* 5(10), 9–24. DOI: 10.56039/rgn10a03.
Katcher, A. and Wilkins, G. (1993) Dialogue with animals: Its nature and culture. *The Biophilia Hypothesis*. Island Press, pp. 173–197.
Kline, C. (2021) Socialising animal-based tourism. In: Higgins-Desbiolles, F., Doering, A. and Chew Bigby, B. (eds) *Socialising Tourism: Rethinking Tourism for Social and Ecological Justice*. Routledge, London and New York, pp. 195–213. DOI: 10.4324/9781003164616.
Lei, S. and Law, R. (2015) Content analysis of Tripadvisor reviews on restaurants: a case study of Macau. *Journal of Tourism* 16(1).
López, Á. and Quintero, G.J. (2021) La geografía del turismo y la geografía de los animales intersectadas por la ética poshumanista. *Cuadernos de Geografía* 30(1), 86–105. DOI: 10.15446/rcdg.v30n1.82631.
Madrid, N. (2004) Sobre el uso de principios teleológicos en la filosofía. *Logos. Anales del Seminario de Metafísica* 37, 7–47.
Méndez, E. and Rodríguez, S. (2013) Definiendo 'lo Mexicano'. Una clave: persistencias del modelo urbano colonial en los 'Pueblos Mágicos.' *Diálogos Latinoamericanos* 21, 46–67.
Mendoza, T. (2015) Importancia ecológica y cultural de una especie endémica de ajolote (*Ambystoma dumerilii*) del lago de Pátzcuaro, Michoacán. *Etnobiología* 10(2), 40–49.
Mitrasinovic, M. (2016) *Total Landscape, Theme Parks, Public Space*. Routledge, London. DOI: 10.4324/9781315236001.
Parra, G., Flores, O. and Mendoza, C. (2014) Biodiversidad de anfibios en México. *Revista Mexicana de Biodiversidad* 85, 460–466. DOI: 10.7550/rmb.32027.
Puebla Noticias (2017) Se prepara el 2o festival de la ajoloteada en Chignahuapan. Puebla Noticias. Available at: http://www.pueblanoticias.com.mx/noticia/se-prepara-el-2o-festival-de-la-ajoloteada-en-chignahuapan-105109 (accessed 10 August 2022).
Quintero, G. and López, Á. (2018) Cambios en la imagen urbana de Chignahuapan, México, a partir de la apropiación del ajolote como elemento turístico patrimonial. In: Alvarado, I. and López, Á. (eds) *Turismo, Patrimonio y Representaciones Espaciales*. PASOS, Tenerife, Spain, pp. 205–230.

Ryder, R.D. (1989) Animal revolution: changing attitudes towards speciesism. *Animal Welfare* 10(2), 222. DOI: 10.1017/S0962728600024003.

Ruiz, J. (2012) *Metodología de la Investigación Cualitativa*. Universidad de Deusto, Bilbao, Spain.

Quintero, G. (2021) El turismo termal y la fabricación de esferas navideñas en Chignahuapan, Puebla como mecanismo de la descampesinización local. *Desarrollo, Economía y Sociedad* 10(1), 26–43. DOI: 10.38017/23228040.737.

Quintero, G. and Castro, A. (2022) La creación del branding turístico y de la sinécdoque turística de Zacatlán a partir de la descampesinización local y el comercio de los productos de manzana. *Revista Ayana* 2(2), 1–17. DOI: 10.24215/27186717e020.

Ríos, S. (2018) El turismo rural de ex-haciendas en Puebla y Tlaxcala y su vinculación a procesos de desarrollo local-regional. *Regiones y Desarrollo Sustentable* 23–24, 53–75.

Rojo, S. and Llanes, R. (2009) Patrimonio y turismo: el caso del programa Pueblos Mágicos. *Revista de Arquitectura, Urbanismo y Ciencias Sociales* 1(3).

Salas, C. and López, Á. (2019) Efectos espaciales de la tematización cultural para la recreación y el turismo en los corredores culturales peatonales del Centro Histórico de la ciudad de México. *Investigaciones Geográficas*. DOI: 10.14350/rig.59763.

Singer, P. (2009) Speciesism and moral status. *Metaphilosophy* 40(3–4), 567–581. DOI: 10.1111/j.1467-9973.2009.01608.x.

Smith, S. (2018) Instagram abroad: performance, consumption and colonial narrative in tourism. *Postcolonial Studies* 21(2), 172–191. DOI: 10.1080/13688790.2018.1461173.

Stephan, E. and Ensástigue, J. (2001) El ajolote, otro regalo de México al mundo. Conabio. *Biodiversitas* 35, 7–11.

Tedlock, B. (1991) From participant observation to the observation of participation: the emergence of narrative ethnography. *Journal of Anthropological Research* 47(1), 69–94. DOI: 10.1086/jar.47.1.3630581.

Tompkins, R. (1978) Genie control of axolotl metamorphosis. *American Zoologist* 18(2), 313–319. DOI: 10.1093/icb/18.2.313.

Urbanik, J. (2012) *Placing Animals: an Introduction to the Geography of Human-Animal Relations*. Rowman & Littlefield, Lanham, Maryland.

Vázquez, A. (2011) El ajolote de Xochimilco. *Ciencias* 98, 098.

Velázquez, M. (2012) Los imaginarios del desarrollo turístico: el programa Pueblos Mágicos en ciudades y comunidades pequeñas de México. *Revista de Arquitectura, Urbanismo y Ciencias Sociales* 3(2), 1–23.

Wolfe, C. (2010) *What is Posthumanism?* University of Minnesota Press, Minneapolis, Minnesota.

11 Feline Saviours: Nekonomics, Stray Agency and the Kawaiification of Heritage Tourism in Japan

Émilie Crossley* and Johan R. Edelheim
Hokkaido University, Hokkaido, Japan

Abstract
This chapter explores the commodification of domestic cats as living mascots at two Japanese heritage tourist attractions: Kishi Station and Bitchū Matsuyama Castle. In both cases, the adoption of a formerly stray cat created a significant boost in tourist numbers at a time of financial hardship, which is interpreted as an example of 'nekonomics' (the economic effect of cats). The cats' presence led to a discursive and material reframing of the heritage sites as *kawaii* (cute) tourist attractions. A posthumanist conceptual lens is used to analyse the cats' disruptive stray agency and the ethics of keeping non-human animals as living mascots in heritage tourism.

Introduction

Japan has a longstanding love of the domestic cat (*Felis catus*). As symbols of good fortune and companions in many homes, cats are extensively represented in artisanal crafts, shrines and popular culture, thus constituting part of Japan's cultural heritage. The economic contribution of this national obsession is captured in the term 'nekonomics' (Miyamoto, 2016). Nekonomics is a Japanese play on words – '*neko*' being the Japanese word for cats and '-nomics' coming from former prime minister Abe's economic policies that were referred to as 'Abenomics'. Nekonomics is increasingly playing a role in tourism, not only through the use of cat imagery at attractions but also through the commodification of live cats, who are often scripted into the role of 'workers'. Little is known about the welfare conditions or experience of these cats, making it challenging to evaluate whether their assimilation into the tourism system constitutes exploitation or a form of multispecies symbiosis.

In this chapter, we analyse nekonomics in heritage tourism through two case studies: Tama the stationmaster of Kishi Station and Sanjūrō the lord of Bitchū Matsuyama Castle. In both cases, the presence of a cat on the premises at a time of financial hardship resulted in a significant boost to visitor numbers. After being taken in by staff, Tama and Sanjūrō were given anthropomorphic outfits to wear and honorary titles. Their popularity has led to the production of cat-themed merchandise and, in Tama's case, Kishi Station was even rebuilt in the shape of a cat's head in 2010. These tourism settings can be conceptualized as sites of emergent, negotiated human–feline relations in which cats actively, if unwittingly, co-create heritage. In the case of Bitchū Matsuyama Castle, this conventional heritage tourist attraction has been modified, in both a narrative and physical sense, by the

*Corresponding author: emilie.crossley@elms.hokudai.ac.jp

 Tourism, Heritage and Commodification of Non-human Animals (eds A. López López *et al.*)
DOI: 10.1079/9781800623309.0011

addition of its cat lord. Kishi Station is more complex; here, we argue that the appointment of Tama and subsequent rebuild of the train station in her image can be understood as the creation of a new heritage tourist attraction, given that the old building was not considered as such. In both contexts, the tangible presence of a living cat has reframed each site as, at least in part, a *kawaii* (cute) attraction.

Tama and Sanjūrō's liminal status as stray or precariously housed cats is significant for how their commodification has been narrativized. Both cats were at the heritage sites of their own volition and entered into seemingly reciprocal relationships with new people; the cats were then provided with shelter, food and care, while in return humans reaped the financial benefits of their presence. The anthropomorphic language used to describe each cat's role has also sought to bolster the legitimacy of their use as living mascots. Tama's 'career' saw her ascend through the ranks to eventually become the Honorary President of Wakayama Electric Rail, while Sanjūrō has been referred to as the 'lord' and 'owner' of his castle. Drawing on a posthumanist theorization, we understand cats as sentient, agentive beings whose lives are entangled with those of humans. We trace the transition in Tama and Sanjūrō's status from liminal to owned cats and the concomitant changes to their agency. Their respective appointments as stationmaster and castle lord undoubtedly resulted in greater material security for the cats. However, this came at the expense of restrictions to their mobility, possibly unwelcome handling by strangers and being made to wear clothes, which we evaluate from both animal welfare and posthumanist ethics perspectives.

The success of the original 'feline saviour', Tama, has inspired copycats around Japan, of which Sanjūrō is just one example. Indeed, the tactic of luring tourists to train stations in particular through the appointment of animal stationmasters extends well beyond the feline realm, with dogs, rabbits, goats, foxes and even lobsters having held the role. It is therefore likely that we will see further 'kawaiification' of Japanese heritage tourism using commodified, living non-human animals in the future. While endorsing a culturally sensitive approach that respects multispecies livelihoods (Thomsen *et al.*, 2021), we also call for greater appreciation of the needs and desires of domesticated animals in tourism who are often overlooked in favour of research on their wild counterparts (Young and Carr, 2018). While it is true that domesticated animals are more likely to thrive in the presence of humans, this does not preclude mistreatment and exploitation from occurring, particularly in heavily commodified tourism settings. In sharing Tama and Sanjūrō's stories, and interpreting them through a posthumanist theoretical lens, we hope to bring nuanced discussion to contemporary ethical debates regarding the roles and treatment of cats in multispecies heritage tourism contexts.

Nekonomics, *Kawaii* Culture and Feline Commodification

In 2016, at the time that the term 'nekonomics' was coined, Japan was going through an unprecedented 'cat boom' with domestic cats overtaking dogs as the nation's companion animal of choice. Subsequently, cats became dominant figures in popular culture, advertising, merchandise and media (Edelheim, 2021). Nekonomics is a concept that thus signifies the economic impact of cat-related consumption in Japan (Miyamoto, 2016). This consumption can be broadly divided into three categories: products and services that meet the needs of pet cats and their owners, such as cat food, toys and veterinary care; products bearing the image or form of cats, such as the Hello Kitty cartoons and associated merchandise; and commodified encounters with living cats in tourism, recreation or hospitality settings, such as cat cafés. It is this latter manifestation of nekonomics that is the main focus of our discussion in this chapter, although the commodification of live cats in tourism is inevitably intertwined with the production of cat-themed imagery in the form of digital and physical signage at tourist attractions and, of course, souvenirs.

Cats are universally recognized as cute animals, which is part of their appeal as companion animals. In Japan, this connection between cats and cuteness is modulated by the pervasive cultural phenomenon of *kawaii*, which is usually translated into English as 'cute', 'loveable' or 'adorable', and describes a range of cultural artefacts (Crossley, 2020). Cats are thought by many people in Japan to be the epitome of *kawaii*, with their neotenic features, soft fur and gentle disposition, which has led to the emergence of various cat-themed *kawaii* brands such as Hello Kitty and Pusheen (Yano, 2013). *Kawaii* animals and objects are believed to have healing (*iyashi*) properties and, indeed, Miyamoto (2021) has recently written about the connection between nekonomics and the healing provided by companion animals. Cat cafés are popular hospitality settings that provide *iyashi* experiences, largely catering for customers who are unable to keep their own pets due to financial, lifestyle, accommodation or allergy issues (Plourde, 2014). These cafés represent one of the most visible forms of contemporary feline commodification and they provide an important cultural backdrop to our chosen heritage tourism settings.

The relationship between *kawaii* and heritage tourism is, perhaps understandably, one that has received little academic attention because it appears to represent such a rare configuration. However, Lee and Yan (2021) argue that cuteness is an important aesthetic in East Asia and that tourists from this part of the world may perceive certain heritage buildings as being cute. Another rare study is Ong's (2021) analysis of a Hello Kitty-themed World Heritage exhibition that took place in 2006 in Macao. He identifies this exhibition as an example of 'cuteified heritage' – a postmodern form of edutainment that uses the *kawaii* aesthetic to communicate heritage values in a physical setting that is removed from the real heritage sites it represents. Ong argues that 'Hello Kitty and Sanrio motifs help subvert the UNESCO expert sanctioned ideals of a heritage place – authentic, academic and serious' (2021, p. 245) and that, in the process, heritage values can be communicated to a broader range of tourists. Our study builds on Ong's work by exploring cuteification within heritage tourism sites themselves. We argue that the commodification of cats at heritage tourist attractions can, as described by Ong, broaden the appeal of such sites and provide novel tools for communicating heritage values. However, a key difference is that the on-site presence of a cute cat and concomitant cat-themed representations changes the nature of the attraction itself and even has the potential to create new heritage tourist attractions.

In the context of living cats, Zhang *et al.* (2021) investigate how stray cats in China's Forbidden City, which is a listed World Cultural Heritage Site, have recently become a popular tourist attraction. Given that the Forbidden City charges an entrance fee, one could conclude that encounters with these cats have become commodified. Furthermore, a range of cat-themed souvenirs are also available to purchase at the site. Zhang *et al.* observe that many domestic tourists visit this heritage attraction with the main purpose of seeing and interacting with the cats, and that they experience enhanced wellbeing as a result. This finding resonates with the healing effects reported by cat café clientele and, although the authors do not explore cuteness as a dimension of the visitor experience, it is almost certain that there is a connection between the cats as cute animals and improved human wellbeing. Similar to the Japanese cases that we will shortly discuss, managers at the Forbidden City attempted to improve public relations following an attack on a British tourist by one of the cats in the 1990s by anthropomorphically referring to them as 'special "security officers" for mice catching in the palace' (Zhang *et al.*, 2021, p. 6). This discursive positioning of the cats as part of the attraction's workforce provided legitimacy to their presence and also potentially enhanced their cute appeal in the eyes of tourists.

Cats, Liminality and Stray Agency

In the context of ongoing animal welfare debates and concerns regarding exploitative practices, a growing number of tourism scholars have sensed the need to include non-human animals in their research. Not content with producing accounts that only foreground tourist and stakeholder voices, many researchers are now contemplating the theoretical and methodological implications of conceptualizing other animals as sentient, agentive subjects. Cohen argues that 'concentration on the human "point of view" and disregard for "the animal point of view" ... is a deficiency that imposes substantial limitations on the insights and conclusions emerging from research into the dynamics of tourist–animal relations' (2020, p. 245). Posthumanism is a philosophical perspective underpinning much of this emerging scholarship, which advocates non-anthropocentrism and understands human and non-human animals as inherently entangled and co-constituting (Hamilton and Taylor, 2017; Guia, 2021). New posthumanist research is therefore inspired by the ethical potential of foregrounding non-human sentience, agency and experience. It is hoped that a deeper understanding of non-human animals may guide best practice, foster empathy in tourists and tourism stakeholders, and lead eventually to more responsible forms of tourism. Such ethical consideration for non-human animals through research in heritage tourism contexts is still in its inception (e.g. Quintero Venegas and López-López, 2021).

Our theoretical approach draws on posthumanism by conceptualizing cats as sentient agents whose lives hold intrinsic value while also being meaningfully entangled with human society. We draw on Donaldson and Kymlicka's definition of agency as 'self-willed or initiated action which carries an expectation of efficacy' (2016, p. 235), which conceptualizes agency as a situated, relational capacity to act independently and intentionally. These authors also provide a useful distinction between what they call 'micro-agency', involving decisions about what to do in the present moment, and 'macro-agency', involving more significant life decisions. One of the consequences for political and ethical theory of understanding other animals as agents is that we can no longer portray them as passive victims of human oppression. As Dashper argues:

> animals are not simply passive victims of human–animal relations. ... Animals can choose to interact with humans or ignore them, to act or not to act in different situations, but these choices are bound by the human-centric context in which these interactions take place. As a result, agency is always relational, contextual and variable and animal agency is always limited by animals' historically- and socially-specific positioning in relation to humans.
>
> (2017, p. 210)

The challenge, of course, is to learn how to recognize and interpret these agentive acts in various different species. In the absence of language (or at least language that humans can understand), animal agency can be inferred through observing behaviour. Buller argues that 'it is predominantly through movement and through their embodied actions that animals negotiate with us and with each other. Movement then, becomes more than functionality and physiological causation, more than agency; it becomes communication' (2012, p. 145). This notion of movement as agency and interspecies communication informs our posthumanist analysis of feline actors in heritage tourism.

The socio-historic positioning of cats in relation to humans has been shaped largely by their domestication as companion animals (Serpell, 2000). As pets, domestic cats cohabit with humans and are assumed to enjoy a comfortable, symbiotic existence in which they receive shelter, sustenance and positive social interactions in return for the companionship they provide to their human families. Cats are among the most physically unrestrained of all pets, and those with unsupervised access to outdoor spaces have the potential to exercise macro-agency by abandoning their household and going to live elsewhere. If not quickly adopted into a new human family, these cats become known as 'strays'; when stray cats breed, their kittens will usually be 'feral' because they are not socialized with humans (Yarri and Stober, 2019). Stray and feral cats can be conceptualized as 'liminal' in that they challenge the binary between wild and domesticated animals (Meijer, 2019). As Meijer notes, 'it is often assumed that companion cats are better off than stray or feral cats, and that cats living on the streets should be rescued, neutered and adopted by humans' (2021, p. 287). However, by exercising 'stray agency' (Meijer, 2021) these cats may resist such efforts to be controlled and socialized by humans. We thus foreground stray agency in order to theorize how our feline protagonists exercise agency, autonomy and resistance across domestic and liminal spaces throughout the course of their journeys into their respective heritage tourism settings.

Free-roaming populations of stray and feral cats are common in many parts of Japan and public opinion on them is divided (Uetake *et al.*, 2014). Kito (2018) suggests that in Japanese culture animals can be regarded as either *Uchi* (inside) or *Soto* (outside) beings, with companion animals belonging to the personal space of the home and thus being considered *Uchi*. However, she argues that an animal which transitions from *Uchi* to *Soto* is controlled by the *Soto* space, regardless of its innate properties, and so stray cats and dogs are not regarded with an ethical awareness by Japanese society (Kito, 2018, p. 170). However, this position appears to be contradicted by empirical evidence suggesting that many Japanese people care for stray and feral cats by feeding them (Uetake *et al.*, 2014). Feral and stray cats are also at the centre of tourist destinations such as Japan's 'cat islands' as well as attracting cat-loving tourists to cities like Onomichi in Hiroshima Prefecture, which has a famous 'cat alley' where cats can be seen. In recent years, the welfare of such cats has come increasingly under the spotlight, partly due to concerns about the risk they pose to human health as carriers of zoonoses (Seo *et al.*, 2021). This has resulted in calls for stricter management of free-roaming cats, such as Seo *et al.*'s recommendation that in Onomichi 'friendly cats should be adopted rather than maintained as community cats' (2021, p. 1). Thus, there seems to be growing discomfort with the liminality and stray agency of non-owned cats in Japan.

Cats in Japanese Heritage Tourism

We now turn to our two case studies, Tama the stationmaster of Kishi Station and Sanjūrō the lord of Bitchū Matsuyama Castle, which illustrate nekonomics and kawaiification in Japanese heritage tourism. In telling the stories of these individual cats, we hope to subvert the dominant tendency of tourism research to document non-human animal collectivities rather than individual lives (Wilkinson, 2018). These stories are the product of focused ethnographic visits to the tourist attractions conducted separately by each author in 2020 and 2022 respectively. Observations of the site, human staff, tourists and cats were made, which were recorded via fieldnotes and photographs. As Tama passed away in 2015, observations at Kishi Station focused on her feline successors, Nitama and Yontama. In addition to these ethnographic visits, a qualitative document analysis was conducted to provide background information on the development of each tourist attraction. Despite the technical distinction between stray and feral cats that was drawn in the literature review, it has been difficult to determine which category the cats in our analysis belong to. In some cases, this is because the origin of the cat is unknown. In other cases, written accounts are ambiguous because in Japanese there is often no linguistic distinction made between stray and feral cats; both are simply referred to as '*noraneko*'.

Tama: stationmaster of Kishi Station

Kishi Station is a railway station on the Kishigawa Line in rural Wakayama Prefecture. First opened in 1933, it used to be a rather small and unremarkable station. Although Kishi Station may have been considered by some to be a heritage building and attracted visiting train enthusiasts, it was not originally a heritage tourist attraction of any significance. However, the fate of the station began to change in 2004 when a financial crisis on the rail line, due to low passenger numbers, threatened it with closure. Tama had been born to a stray cat called Miko and was raised by a shopkeeper at Kishi Station along with another stray cat, Chibi. Interestingly, however, both Japanese and English media frequently refer to Tama as a stray cat. As a result of the financial crisis, a review of Kishi Station's maintenance policy was conducted and there was a possibility that the cats would not be permitted to remain at the shop where they were living. As a solution to this dilemma, the shopkeeper who took care of the cats and did not want to see them moved proposed to Mitsunobu Kojima, the president of Wakayama Electric Railway, that Tama should be appointed as a feline 'stationmaster' and, in 2007, this appointment took place (Nonaka, 2021). Miko and Chibi were appointed as Tama's assistants, so they too were able to stay.

Tama's appointment as stationmaster marked a turning point for Kishi Station. Visitors from all over Japan began to arrive in Wakayama to meet this famous cat, who was occasionally made to wear a cute, anthropomorphic uniform

Fig. 11.1. Tama Museum Kishi Station. (Photo by author.)

that included a custom-made hat and name badge that hung from her collar. A lighter version of the hat was even crafted for the summer months to prevent Tama from overheating. Railway passenger numbers increased to such a degree that both the line and station were saved from closure. In 2009, a new 'Tama Train' began to operate on the line, which was decorated inside and out with cute cartoon depictions of Tama. A year later, the main building of Kishi Station, which had been deteriorating, was rebuilt in the shape of a cat's head with the word 'TAMA' placed on its ridge. In its new form, the station is now referred to as the 'Tama Museum Kishi Station' (Fig. 11.1). Inside, a new seating area, café and gift shop containing Tama-themed merchandise were created in order to provide a more comprehensive visitor experience for tourists. The *kawaii* aesthetic infuses the whole experience, from the cartoon version of Tama in sign-form greeting visitors at the entrance, to the cat-shaped desserts available at the café, to the anthropomorphic Tama Train with whiskers painted on its 'face' (front end) (Fig. 11.2).

Tama's 'career' saw her ascend through the ranks to eventually become the Honorary President of Wakayama Electric Rail before her death in 2015 at the age of 16 years, after which she was enshrined at the station and deified. Today, Tama has two successors, Nitama (Tama II) and Yontama (Tama IV), who are also former stray female calico cats, thus evoking the image and spirit of the original feline stationmaster (Videsjorden, 2020, p. 38). Currently, both cats 'work' 6 hours a day, 5 days a week, during which time they are confined to glass cages. Nitama spends her 'shifts' at Kishi Station, while Yontama's 'shifts' are split between Kishi and Idakiso Station further along the line (Figs 11.3 and 11.4). The cages are about one-metre wide and deep, with a litter tray at the bottom, a cat bed and several ledges leading upwards to the top of the cage, which extends to the ceiling. On one of the ledges in each cage is a tiny stationmaster hat. The cage at Kishi Station

Fig. 11.2. The Tama Train. (Photo by author.)

opens on one side to the gift shop, allowing staff to reach in and pet the cat with ease, whereas the cage at Idakiso Station is completely sealed. Signs inform visitors in Japanese, Chinese and English not to use flash photography or beat on the windowpanes. However, we observed several people lightly tapping on the glass who were not stopped by staff. Unlike the early days of Tama's residence at the station, visitors are now unable to touch the stationmaster cats and must content themselves with taking photographs of them instead.

It is estimated that Tama's presence contributed ¥1.1 billion to the local economy during her years of active service at Kishi Station, making her a true nekonomic success story that other railway stations and tourist attractions in Japan have tried to emulate (Videsjorden, 2020). During the COVID-19 pandemic, tourist numbers dropped significantly, threatening the station financially once again. In 2021, the railway's management launched a newly designed vehicle called the 'Tama Train Museum' amid much publicity in the hope of once again reaping the nekonomic rewards of their feline saviour (Nonaka, 2021). In terms of its connection to heritage tourism, Kishi Station has been physically and representationally transformed from an inconspicuous rail heritage building into a brand new tourist attraction that combines rail heritage and the *iyashi* (healing) power of cute cats. Furthermore, we can conceptualize the 'Tama Museum Kishi Station' as the creation of a new, *kawaii* cultural heritage site in which the legacy of Tama, who is now so much a part of the local community's story, is commemorated. However, further down the line at Idakiso Station, recently installed information panels documenting various historically significant rail structures make no mention of the stationmaster cats, thus signalling a divergence between traditional and kawaiified heritage narratives.

An important part of the narrative surrounding Tama is that she was a cat already residing at Kishi Station; she was not purchased

Fig. 11.3. Nitama's cage at Kishi Station. (Photo by author.)

or brought in from elsewhere against her will. Kishi Station was already Tama's home and her apparent lack of resistance to wearing a 'uniform' or performing 'duties' gave the impression that her appointment as feline stationmaster was consensual in nature. Her transition from being precariously housed to having the material securities afforded by being a stationmaster framed Tama's appointment as a symbiotic relationship benefiting both the feline and human parties. However, this transition came with certain costs as well, including restrictions to her mobility and the inconvenience of having to wear a hat, even if it was designed to be very lightweight. In her first few years as feline stationmaster, Tama was kept in her 'office' (a glass cage) but was also allowed to sit on a cushion outside the ticket office for visitors to inspect her more closely and have photos taken with her. However, as the station grew in popularity, management of the cats became stricter and nowadays Nitama and Yontama appear to remain in their cages during 'work' hours.

Sanjūrō: lord of Bitchū Matsuyama Castle

Bitchū Matsuyama Castle in Takahashi City of Okayama Prefecture is also called the 'castle in the sky' as it appears to be floating among clouds on autumn and winter mornings when clouds cover the surrounding valleys. It was originally built in 1240 and is counted as one of Japan's more prominent castles, especially as it is one of the few that was not rebuilt with a concrete base and structure after the Second World War, but rather refurbished in an original style using local timber. The castle sits atop a 430-metre-high hill a 20 minute drive or an hour's walk away from Takahashi City. Due to its elevated location, it played an important role in the warring period ahead of the Edo era as it was easy to protect and its location gave its owners a good view of the surrounding landscape. This became, however, also its major disadvantage in the peaceful time that followed, and by early Showa era it was mostly just in ruins. It was thanks to

Fig. 11.4. Yontama's cage at Idakiso Station. (Photo by author.)

a major restoration effort by townspeople that it was returned to its former glory (Nishimatsu, 2019).

Bitchū Matsuyama gained in popularity as a heritage tourist attraction during the 2000s, often with the help of different media linkages. Pictures of the 'castle in the sky' were linked to anime films, and in 2016 images of it featured in the opening scenes of *Sanada Maru*, NHK's annual 'Taiga Drama' (a 50-episode-long historical drama shown once a week by the national broadcaster on prime time television). Visitor numbers rose 1000-fold during the 2000s, and the annual visitor count in 2016 was 106,000. The road leading up to the castle was, however, severely damaged by torrential rain in 2018, making it impossible for heavier vehicles to get up to the previous turning point, and even for the castle's service vehicles to make it to the castle. This had a drastic impact on visitor flows to the castle, and could have been disastrous for the attraction had it not been for a cat stepping in as its 'saviour'.

The same heavy rain that damaged the roads up to the castle also caused flooding in Takahashi City, and one resident who apparently decided to relocate for higher ground was a yellow and white tabby cat. Caretakers found the cat when they arrived on foot to the castle a few weeks after the torrential rain. This was surprising as no humans live in the vicinity and it is a strenuous walk for anybody to get up there. The cat seemed to have taken a liking to the surroundings and, despite being free to leave, he stayed put in the castle. Visitors started returning when they heard about the friendly cat who greeted visitors to 'his' castle. This became a media hit in Japan with major newspapers and television stations reporting on the cat taking up residence in the castle. The local tourism association decided therefore to officially proclaim him the castle lord, giving him the name Sanjūrō in reference to a local samurai family that had served as warriors defending Osaka in the warring period (Nishimatsu, 2019). The large influx of visitors who came to see Sanjūrō became the catalyst for the tourism association to repair all access roads, and visitor numbers have since surpassed the records set before the castle got its feline lord.

Sanjūrō's journey to becoming a castle lord was, however, rather fraught. Shortly after the initial wave of publicity, a couple arrived at the castle claiming that Sanjūrō was in fact their cat called 'Natsume' (Nishimatsu, 2019). They had taken him in as a stray but he had seemed unsettled at their house and escaped from the balcony after only a short time. Seeing how much more settled the newly named Sanjūrō looked at Bitchū Matsuyama Castle, the couple decided that it was in his best interests to remain there. More drama unfolded when Sanjūrō disappeared for 2 days after an unwelcome trip to the vets and then, not long afterwards, ran away for nearly 3 weeks. When he was eventually found, Sanjūrō had lost a significant amount of weight and the castle management had been inconvenienced by the loss of their new, prized asset (Nishimatsu, 2019). A plan was thus devised to establish a mutually beneficial arrangement whereby Sanjūrō would be contained within the castle grounds – his stray agency was perceived as a problem.

Under this new arrangement, Sanjūrō is kept in a large cage within the castle at night and during the day he has a routine that consists of two meals, two supervised walks around the castle grounds, and two sessions during which tourists can photograph and interact with him. In order to prevent him from running away again, Sanjūrō is now taken for walks on a leash and harness (which he does not seem to mind) by one of his 'human servants' (Nishimatsu, 2019, p. 109). However, during one of our visits to Bitchū Matsuyama, visitors were told that Sanjūrō would not be taking his morning walk since it was too hot that day. Instead, he slept and groomed in a covered bed in the castle courtyard while being guarded by a member of staff (Fig. 11.5). This indicates that Sanjūrō's routine is not set in stone, and that his welfare is prioritized over tourists' desires to see him perform. In addition to his walks, Sanjūrō can also wander around the castle courtyard while on a long leash. On occasions, Sanjūrō has been seen dressed in tiny, custom-made samurai outfits, the images of which have led to viral popularity on social media. However, it appears that Sanjūrō only wears these outfits for very short periods of time during promotional photoshoots and that most of the time he only has a collar on (Fig. 11.6).

Sanjūrō's photographic and cartoon image is used on souvenirs, such as keyrings, stickers and biscuits. There is even a Sanjūrō-themed taxi in Takahashi City! At the castle itself, cartoon Sanjūrō can be found acting as a cute guide on various signs asking visitors to comply with COVID-19 social distancing rules. It is clear that he is a recent addition to the castle, as all of the permanent displays focus exclusively on the castle's history as a heritage building rather than on the presence of its feline castle lord. In February 2022, a stone bench bearing a sculpture of Sanjūrō sitting on a cushion was installed in front of the castle for tourists to use as a photography spot (Fig. 11.7). The bench can be interpreted as a physical change to the heritage site brought about by a process of nekonomic kawaiification that embodies and commemorates the latest phase of Bitchū Matsuyama Castle's evolution.

Discussion

Japan's feline stationmasters and castle lord represent the alignment of usually separate cultural elements: heritage and *kawaii*. Although these living mascots may seem like a very niche form of animal-related tourism, they can be interpreted as manifestations of the broader phenomenon of nekonomics, which has proven to be a powerful socio-economic force in Japan (Miyamoto, 2016). As our case studies reveal, the presence of a cute, anthropomorphically framed cat at heritage sites can diversify and grow the attraction's visitor base, leading to significant financial rewards. Given that this is the case, we predict that there will be further kawaiification of Japanese heritage tourism in the future. However, beyond this economic impact, kawaiification also appears to entail a discursive and even material reimagining of heritage. Rather than simply being gimmicks that have no modifying effect on the heritage tourist attraction within which they are situated, Sanjūrō, Tama and her successors have become integrated with the castle and

Fig. 11.5. Sanjūrō on a leash in his day-bed in front of Bitchū Matsuyama Castle. (Photo by author.)

rail heritage narratives at their respective sites. This entanglement has been further embedded through physical alterations to the sites, such as the extensive cat-themed rebuild of Kishi Station or the more modest addition of the new cat-sculpture bench at Bitchū Matsuyama Castle. While Ong (2021) has expressed support for 'cuteification' strategies that can broaden the appeal of heritage, for some the physical alteration of heritage sites may prove to be a step too far if it is perceived as eroding their authenticity or damaging historically significant structures.

Fig. 11.6. Photobook at Bitchū Matsuyama Castle showing images of Sanjūrō from 2020. (Photo by author.)

The successful integration of our feline saviours into the tourism system was facilitated by narratives portraying them as freely choosing to live at the heritage sites (at least initially) and enjoying an enhanced quality of life as a result. The cats' celebrated transition from stray, or precariously housed, to owned reflects a growing sentiment in Japanese society that liminal 'community cats' need to be more carefully managed and controlled (Seo *et al.*, 2021). However, while the notion that Tama and Sanjūrō willingly submitted to their new captivity and 'duties' may be appealing to tourists because it diminishes concerns regarding potential mistreatment, the reality of the situation is more complex. In the case of Sanjūrō, an initial choice to live at Bitchū Matsuyama Castle was quickly followed by several macro-agentive attempts to relocate elsewhere. It is possible that he was searching for other cats with which to socialize or breed. Sanjūrō's stray agency and apparent resistance to being owned were deemed problematic by the castle's staff and, as a result, he is now prevented from leaving the castle grounds (Nishimatsu, 2019). While Tama may not have displayed the same desire to wander, she too had her mobility limited, as did her successors Nitama and Yontama. Effectively, the cats' liminality and stray agency were erased through their integration into the heritage tourist attractions.

It is understandable that the humans caring for these cats may have wanted to restrict their movements due to safety concerns; indeed, a feline stationmaster in Kyushu who enjoyed a greater degree of freedom was recently hit by a car and killed while taking an evening stroll (Baseel, 2021). Sanjūrō was also found to have lost a significant amount of weight during his lengthier escapade, indicating that his freedom came at a cost to his health (Nishimatsu, 2019). There is clearly a tension between the viewpoint that cats, as domesticated animals, require human care

Fig. 11.7. New bench at Bitchū Matsuyama Castle featuring a statue of Sanjūrō. (Photo by author.)

and management, and manifestations of feline stray agency in which cats choose a riskier, less comfortable existence beyond the confines of pet ownership. Agency is one of several posthumanist ethical dimensions that requires consideration when evaluating the practice of keeping living mascots at tourist attractions. In order to meet the expectations of tourists and to function as an attractor, the cats are required to be at the heritage sites at certain times. Beyond appeals to their safety, other rhetorical strategies are used by the tourist attractions to legitimize restrictions to the cats' mobility, such as anthropomorphically positioning the stationmasters as having 'duties' and 'work hours' (as well as days off).

While a more extensive study would be required to ascertain fully the welfare status of these cats, our preliminary observations indicate that they are generally well cared for, supervised by staff at all times and not normally made to wear clothes other than during ceremonial occasions or promotional photoshoots. However, while Sanjūrō is able to spend a significant portion of his day outside and can take walks on a leash, the level of confinement experienced by the feline stationmasters is far more restrictive. On a recent visit to Idakiso Station, Yontama appeared agitated and frustrated; she paced inside her cage, meowed repeatedly and stared intensely at something that had caught her attention on the platform. Some of these behaviours may indicate the development of stereotypies, although further observations would be required to confirm this. The large name tag on Yontama's collar swung from side to side and knocked loudly against the sides of the cage as she walked, which may be irritating for her. Nitama appears unable to use the small ledges inside her cage due to her large size, diminishing the usable space within the cage.

Although both cages appear to meet welfare requirements in terms of dimensions, they may be lacking in terms of enrichment and social contact (Stella *et al.*, 2017). It was also concerning to observe the repeated flouting of rules as visitors tapped on the stationmaster cats' cage windows.

A posthumanist perspective can be used to critique the practice of keeping living mascots at heritage tourist attractions on the basis that the animals' needs and desires are positioned hierarchically below those of humans. It is clear that tourists' desire to see and photograph the cats, and the economic benefits that these tourists bring to other humans in the area, are of primary importance to the operators. It would have been possible, for example, to create a larger, more enriched cage at Kishi Station during its rebuild, but this would have reduced space within the gift shop – thus impeding the station's ability to profit from the feline stationmasters. While many of the cats' basic welfare needs appear to be met, it is possible that their lives are not as fulfilling as they would be if they were allowed to roam more freely and socialize with other cats. However, simply turning these cats loose would almost surely condemn them to greater suffering and a shorter life, while also potentially having a negative impact on native wildlife. We believe that it is possible for mutually beneficial interspecies relationships to occur in tourism but that more needs to be done to understand the needs and desires of domesticated animals to raise them from their current subordinate status.

Conclusion

In conclusion, we have analysed two tourism settings in Japan that can be conceptualized as sites of emergent, negotiated human–feline relations in which cats actively, if unwittingly, co-create kawaiified heritage. The popularity of the cats on the Kishigawa Line and at Bitchū Matsuyama Castle demonstrate how nekonomics can rapidly turn around the fortunes of heritage sites. However, the practice of keeping commodified living mascots at tourist attractions comes with significant ethical responsibilities and practical challenges. While we believe that such arrangements can in some cases be mutually beneficial for both the human and non-human actors involved, our research reveals cause for concern regarding extreme restrictions to the cats' mobility at Kishi and Idakiso Stations. It is possible to create a middle ground between unrestricted stray agency and confinement in tiny cages, which ensures that human enjoyment of living mascots does not come at the expense of the animals' welfare. This appears to have been achieved at Bitchū Matsuyama Castle, where Sanjūrō is available for tourists to view but is also enabled to wander the castle grounds in a safe, supervised manner. However, this arrangement calls into question whether Sanjūrō's agency is being sufficiently respected. The cases we have examined reveal the complex interplay between economics, *kawaii* culture and interspecies relationality, highlighting the need for further posthumanist research into multispecies heritage tourism contexts involving domesticated animals.

Acknowledgements

This research was generously supported by a JSPS Postdoctoral Fellowship for Research in Japan and JSPS KAKENHI: Grant Number 22F20797.

References

Baseel, C. (2021) Stationmaster cat dies in Japan after being hit by car. SoraNews24. Available at: https://soranews24.com/2021/06/12/stationmaster-cat-dies-in-japan-after-being-hit-by-car/ (accessed 25 May 2022).

Buller, H. (2012) "One slash of light, then gone". Animals as movement. *Études Rurales* 189, 139–153. DOI: 10.4000/etudesrurales.9630.

Cohen, E. (2020) Rethinking the animal in tourism studies. In: Dieke, P.U.C., King, B.E.M. and Sharpley, R.A.J. (eds) *Tourism in Development: Reflective Essays*. CABI, Wallingford, UK, pp. 245–255. DOI: 10.1079/9781789242812.0000.

Crossley, É. (2020) Japanese animal tourism and the kawaii (cute) aesthetic. *Scope (Art & Design)* 19, 65–71. DOI: 10.34074/scop.1019013.

Dashper, K. (2017) Listening to horses: developing attentive interspecies relationships through sport and leisure. *Society and Animals* 25(3), 207–224. DOI: 10.1163/15685306-12341426.

Donaldson, S. and Kymlicka, W. (2016) Comment: between wild and domesticated: rethinking categories and boundaries in response to animal agency. In: Bovenkerk, B. and Keulartz, J. (eds) *Animal Ethics in the Age of Humans: Blurring Boundaries in Human-Animal Relationships*. Springer, Cham, Switzerland, pp. 225–242. DOI: 10.1007/978-3-319-44206-8.

Edelheim, J.R. (2021) Nekonomics and feline stationmasters – the point of putting a hat on a cat. In: Trupp, A., Wong, P.P.W., Chaichi, K. and Loh, E.G.Y. (eds) *TRMC 2021 – 3rd International Tourism and Retail Service Management Conference*, Sunway University, Kuala Lumpur, Malaysia, pp. 53–54.

Guia, J. (2021) Conceptualizing justice tourism and the promise of posthumanism. *Journal of Sustainable Tourism* 29(2–3), 503–520. DOI: 10.1080/09669582.2020.1771347.

Hamilton, L. and Taylor, N. (2017) *Ethnography After Humanism: Power, Politics and Method in Multi-Species Research*. Palgrave Macmillan, London. DOI: 10.1057/978-1-137-53933-5.

Kito, Y. (2018) The metaphysical background of animal ethics and tourism in Japan. In: Kline, C. (ed.) *Tourism Experiences and Animal Consumption: Contested Values, Morality and Ethics*. Routledge, Abingdon, UK, pp. 165–178. DOI: 10.4324/9781315265186.

Lee, H. and Yan, L. (2021) Selling cute destinations to East Asia. *Journal of Hospitality and Tourism Insights* 4(3), 282–299. DOI: 10.1108/JHTI-04-2020-0039.

Meijer, E. (2019) *When Animals Speak: Toward an Interspecies Democracy*. New York University Press, New York. DOI: 10.18574/nyu/9781479859351.001.0001.

Meijer, E. (2021) Stray agency and interspecies care: the Amsterdam stray cats and their humans. In: Bovenkerk, B. and Keulartz, J. (eds) *Animals in Our Midst: The Challenges of Co-Existing With Animals in the Anthropocene*. Springer, Cham, Switzerland, pp. 287–300. DOI: 10.1007/978-3-030-63523-7.

Miyamoto, K. (2016) ネコノミクスの経済効果 *[Economic effects of nekonomics]*. Available at: https://www.kansai-u.ac.jp/global/guide/pressrelease/2015/No44.pdf (accessed 2 February 2022).

Miyamoto, K. (2021) コロナ禍の癒し「ネコノミクス」と「イヌノミクス」*[Corona Healing "Nekonomics" and "Inunomics"]*. Available at: https://www.kansai-u.ac.jp/ja/assets/pdf/about/pr/press_release/2020/No65.pdf (accessed 2 February 2022).

Nishimatsu, H. (2019) さんじゅーろ ー, 備中松山城, 猫城主 [*Sanjūrō – Bitchū Matsuyama Cat Castle Lord*]. Kabushikigaisha ha-to shuppan [Heart Publishing].

Nonaka, R. (2021) 「たま駅長」は不滅の存在感,再びの大ピンチ救って ["Stationmaster Tama" has an immortal presence, saving the day once again]. *Asahi Shimbun*. Available at: https://www.asahi.com/articles/ASP7F4W45P71PLBJ002.html (accessed 2 February 2022).

Ong, C.E. (2021) The past is a feline country? Cuteification-value nexus and world heritage visitor engagement. *Tourism Critiques: Practice and Theory,* 2(2), 235–249. DOI: 10.1108/TRC-08-2021-0015.

Plourde, L. (2014) Cat cafés, affective labor, and the healing boom in Japan. *Japanese Studies* 34(2), 115–133. DOI: 10.1080/10371397.2014.928183.

Quintero Venegas, G.J. and López-López, Á. (2021) Working donkeys in northwestern Mexico: urban identity and tourism resources. In: Rickley, J.M. and Kline, C. (eds) *Exploring Non-Human Work in Tourism: From Beasts of Burden to Animal Ambassadors*. De Gruyter, Berlin, pp. 53–68. DOI: 10.1515/9783110664058.

Seo, A., Ueda, Y. and Tanida, H. (2021) Health status of "community cats" living in the tourist area of the old town in Onomichi City, Japan. *Journal of Applied Animal Welfare Science* 25(4), 338–354. DOI: 10.1080/10888705.2021.1874952.

Serpell, J.A. (2000) Domestication and history of the cat. In: Turner, D. and Bateson, P. (eds) *The Domestic Cat: The Biology of Its Behavior*. Cambridge University Press, Cambridge, UK, pp. 83–100. DOI: 10.1017/CBO9781139177177.

Stella, J.L., Croney, C.C. and Buffington, C.T. (2017) Behavior and welfare of domestic cats housed in cages larger than U.S. norm. *Journal of Applied Animal Welfare Science* 20(3), 296–312. DOI: 10.1080/10888705.2017.1317252.

Thomsen, B., Thomsen, J., Copeland, K., Coose, S., Arnold, E. *et al*. (2021) Multispecies livelihoods: a posthumanist approach to wildlife ecotourism that promotes animal ethics. *Journal of Sustainable Tourism* 31(5), 1195–1213. DOI: 10.1080/09669582.2021.1942893.

Uetake, K., Yamada, S., Yano, M. and Tanaka, T. (2014) A survey of attitudes of local citizens of a residential area toward urban stray cats in Japan. *Journal of Applied Animal Welfare Science* 17(2), 172–177. DOI: 10.1080/10888705.2013.798558.

Videsjorden, S. (2020) 'Tama between realms': riding the cat train through rural Japan – a cultural study of Tama the stationmaster. Publication Number (87346) University of Oslo. Available at: https://www.duo.uio.no/handle/10852/84626 (accessed 31 March 2022).

Wilkinson, S. (2018) Being Camilla: the 'leisure' life of a captive chameleon. In: Carr, N. and Young, J. (eds) *Wild Animals and Leisure: Rights and Wellbeing*. Routledge, Abingdon, UK, pp. 96–112. DOI: 10.4324/9781315457413.

Yano, C.R. (2013) *Pink Globalization: Hello Kitty's Trek Across the Pacific*. Duke University Press, Durham, North Carolina. DOI: 10.1515/9780822395881.

Yarri, D. and Stober, S.S. (2019) To intervene or not to intervene? The issue of the liminal feral cat. *Open Journal of Philosophy* 9(2), 204–222. DOI: 10.4236/ojpp.2019.92014.

Young, J. and Carr, N. (2018) Introduction. In: Young, J. and Carr, N. (eds) *Domestic Animals, Humans, and Leisure: Rights, Welfare, and Wellbeing*. Routledge, Abingdon, UK, pp. 1–11. DOI: 10.4324/9781315457451.

Zhang, Y., Lee, T.J. and Xiong, Y. (2021) The contribution of animal-based cultural ecosystem services to tourist well-being at cultural heritage sites. *Journal of Travel Research* 61(7), 1631–1647. DOI: 10.1177/00472875211037743.

12 The Heritage of the Cute: Commodifying Pandas in Urban and Rural China

Yulei Guo*
Chengdu Research Base of Giant Panda Breeding, Chengdu, China

Abstract
In panda tourism, the panda materializes the cute aesthetic in addition to its symbolic connection to wildlife conservation. This chapter examines the panda's cuteness as a significant cultural heritage since the modern industrialization that panda tourism facilitates in both urban and rural China today. The heritage of the cute in panda tourism signifies two things: first, the emergence of an attraction based on emotions and feelings that can be commodified and consumed by tourists through the aestheticization of the charisma and appearance of the cute panda; and second, the provision of a ritual, a structured means, for designing, producing and consuming the heritage of the cute. The chapter examines Chengdu's metropolitan area and Ya'an, a rural area close to Chengdu, as significant sites where the heritage of the cute has been incorporated into landscapes shaped by panda tourism.

Introduction

The giant panda, the logo of the World Wildlife Fund (WWF) since 1961, together with its sanctuary in Sichuan, China, were added to the UNESCO World Heritage List in 2006 (WWF, 2006). The heritage list provides recognition of the extraordinary conservation efforts for the panda and the natural values of the animal's habitat. Serving as a landmark of wildlife conservation, the value of ecosystem services of the giant panda was US$2.6–US$6.9 billion in 2010 (Wei *et al.*, 2018). However, while the panda and its habitat as a natural heritage have been well identified among international organizations, its cultural importance and value, which are rooted in modern and contemporary societies and the commodification of culture, have received less attention. Yet, as the historian Songster writes, 'it can be argued that it [the panda] has survived extinction because it is cute' (2018, p. 173). The cuteness of the giant panda has created a sufficient amount of social capital that has saved the animal from extinction. In contrast to the giant panda as wildlife and the natural habitat the panda lives in, the cuteness of the panda refers to a societal construction that, according to Ngai (2005), emerged as a term of evaluation and a formally identifiable style in the USA in the industrial 19th century. In this chapter, the cuteness of the giant panda is considered a parallel to its natural heritage, serving as a cultural heritage that has been consumed and commodified in social experiences such as tourism.

The understanding of the heritage of the cute belongs to a growing body of posthumanism

*yulei.guo@panda.org.cn

 Tourism, Heritage and Commodification of Non-human Animals (eds A. López López *et al.*)
DOI: 10.1079/9781800623309.0012

studies on animal welfare and ethics in tourism (Fennell, 2013; Carr and Broom, 2018; Kline, 2018; Winter, 2020; Rickly and Kline, 2021). From a posthumanist perspective, the heritage of the cute builds on tourist–panda interactions, drawing on tourists' pursuit of satisfaction (Burns, 2015), transforming pandas into unrecognized workers (Guo, 2021). However, the heritage of the cute, as the chapter aims to show, has been interwoven with the heritage of conservation to the extent that panda conservation efforts are deeply rooted in the consumption and recognition of the heritage of the cute. Based on the theoretical frameworks of affective economics and lively capital, the author has conducted extensive participant observations in both urban and rural China since 2019, where a wealth of field photos, notes, videos and booklets thematically relating to the giant pandas and their cuteness have been accumulated. Through further analysis of this collection, this chapter demonstrates the construction and structure of the heritage of the cute and its inevitable link to conservation efforts. The heritage of the cute provides a conundrum that necessitates research that goes beyond the critical understanding of the unequal connection between pandas and tourists but focuses on its role in supporting and enabling pandas and other animal conservation initiatives.

The heritage of the cute builds on affective economies that demand an intensive exchange of emotional labour between bodies. According to Ahmed (2004), in affective economies, emotions can move beyond the realm of private individuals and are exchanged between bodies, shaping different surfaces or boundaries. Affective economies drive emotions and affect to circulate between tourists and pandas, which help to capitalize on passion 'accumulated over time' (2004, p. 120). The heritage of the cute, as the chapter will show, grows from an intensive exchange of emotional labour between giant pandas and tourists and a consistent accumulation of affects and passion over time. A retrospective look at the panda's integration into human society manifests a process of accumulated passion towards the animal as affects and emotions circulate between the two species. For example, Songster (2018) considers the panda's popularity within China 'largely a post-1949 phenomenon' and hence a by-product of the establishment of the People's Republic of China. Nicholls (2012) contrasts the panda's current recognition with its unknown history before 1869. The heritage of the cute builds on the accumulated affective economics towards the panda over time, showing itself in contemporary China as a landmark of consumption (e.g. the panda sculpture on Chengdu International Finance Square) and conservation (e.g. the panda in roadside bushes in Fazhan Village).

In her study of the digital media economy, Lukacs (2015) contextualizes the consumption and production of the cute through affective economies among Japanese young women, showing that, as humans, the Japanese young female can leverage their cute labour to gain upward social mobility. While Japanese female internet idols have purposefully participated in the cute labour, from a posthumanist perspective, we are obliged to consider whether pandas have ever reached the consensus of becoming an embodiment of the heritage of the cute. Dale (2016) argues that non-human species, such as the Siberian silver fox, perform self-domestication as people select cute foxes to tame. For Dale (2016), cuteness can pave the way to inter- or intra-species coexistence. Pandas, however, have rejected the self-domestication agenda from the beginning with their reluctance to breed in captivity (Diamond, 2002). This study shows that affective economies concerning the cute have preconditioned the panda's 'rebel', which has further contributed to the animals' iconic celebrity status in the animal kingdom and made it a flagship species for wildlife conservation. In other words, the heritage of the cute established on the giant panda has been a human social project that has harnessed what Haraway (2013) calls the 'lively capital' within the animal that has both enabled the capitalization of the animal's cute appearance as well as the animal's reluctance to breed.

This chapter anchors the investigation of the heritage of the cute in panda tourism, where tourists encounter pandas as lively capitals in zoos or breeding centres. It argues that panda tourism provides a vehicle for emotions and affects exchanges between tourists and the pandas, and while pandas materialized into lively capitals, an unequal and anthropocentric relationship has dominated the encountering experiences between tourists and pandas.

Building on Ahmed's theory of affective economics and Haraway's concept of lively capital, the chapter brings a lengthy ethnographic study in urban and rural China into the picture, showing that the heritage of the cute has been integrated into the everyday lives of tourists and urbanites for both consumption and conservation in contemporary China.

Affective Economies of Cute

In his epilogue to *The Way of the Panda*, Nicholls (2012) concluded that WWF's application of the panda as the organization's logo has created a turning point that has transformed the real panda into a virtual animal. Nicholls (2012) furthers that, like the construction of a successful brand, the WWF panda has become shorthand for an array of positive associations, including charisma, wilderness, biodiversity, philanthropy and a promising future. What Nicholls has observed in his tracking of the rise of the image of the giant panda since its discovery is a series of historical moments when the affective economies of the giant panda were asserted, affirmed and further produced and reproduced through the popularization of the image of the wildlife. Despite Nicholls' close investigation of the virtualization of panda into one of the most successful brands in the world, the mechanism with which the branding was possible has been left largely unexamined. However, it is necessary to ask why and how certain positive emotions are bound to the giant pandas, transforming an unknown animal 150 years ago into the brightest animal celebrity in contemporary society.

The notion of affective economies was first introduced by Sara Ahmed (2004), a scholar who investigated the construction of a white subject and national identity based on hate as a collective emotional experience among citizens. She identified three components of an affective economy: the creation of a mutual identity, the capitalization of an emotion exchange process and the fetishistic differentiation inevitably attached to the process of capitalization. Ahmed (2004) writes that the experience of hate creates a white subject collectively shared among citizens. The collective identity is further enhanced as subjects circulate the same hate experiences between each other, despite the fact that hate does not reside with the white subject and the identity initially. Ahmed (2004) explains that once identified with the collective identity, a boundary between 'them' and 'us' is generated, enforcing movements that further strengthen emotional attachment and affection. Since Descartes, the concept of otherness has been deemed inferior and peripheral in contrast to the privileged centre of individual consciousness. The affective economics in contemporary society reinforces this other–us gap by capitalizing on emotions and affect that are further embodied in individuals.

While Ahmed's study of affective economies aims at explicating issues surrounding white subjects and others in Western countries, the making of the heritage of the cute, I argue, follows a similar model of emotional 'operation' (Ahmed, 2004, p. 124) that, despite its positive association with signs such as wilderness, loveliness and philanthropy, has contributed to the differentiation and, moreover, an unequal relationship between the panda and tourists. First, cuteness generates an experience of attachment that binds tourists together. As Yano has observed in her study of the globalized Hello Kitty's cute images, the cat 'leads the way in this global process of identity making through (often underage) consumption' (2013, p. 31). The image of the cute panda demands affects and emotions from tourists who are mutually connected to one another, as the same emotions circulate among the tourists. The desired movement between the tourist subject and the panda as the object is the consumption of the giant pandas and their cute images. Scholars believe that cuteness contributes to an aesthetic category along with the rise and dominance of consumerism (Harris, 2000; Genosko, 2005; Ngai, 2005). Categorizing the giant panda into the aestheticism of the cute displaces the animal from the natural habitat it lives in and transforms the wildlife into a commodity that modern tourists can consume. The circulation of affects surrounding the cute giant pandas accumulates into a capital that continues to appreciate to the extent that the heritage of the cute emerges from the monumental increase.

On the surface, the consumption of cuteness has brought unexpected benefits to the animal. In Songster's (2018) words, the giant panda is thus saved from its fate of extinction for the sake of its cuteness. The giant panda becomes a spokesperson for the kindness and benevolence of humans. However, like Ahmed's observation of the hate emotions attached to racism, such positive affects associated with the cute giant pandas operate in a similar manner in which the giant pandas have been aligned with 'the other'. As Genosko observes, the connection with the cute allows people to become nurturers, breeders, buyers and judges, who share 'delight and emotional warmth' with the commodity but are 'not understanding and respecting' (2005, np). Displacing the giant pandas from their natural habitat enables a relationship of nurturing and breeding between the tourists and the pandas, but such a relationship only encourages the nurturers' consumption rather than a better understanding and respect for the animal. Haraway's (2013) concept of 'lively capital' helps us better understand visitors' commodified experiences of the giant pandas.

Pandas as Lively Capitals: The Construction of the Heritage of the Cute

Affective economies provide a framework within which the tourist's affect and experiences with the pandas have been contextualized. However, the framework has left the participation of the pandas aside. The fact that pandas conduct the labour of paws rather than hands, in Haraway's (2013) terms, suggests that pandas' participation in the making of the heritage of the cute is not purposeful or emotional. Rather, the pandas are best understood as 'performers' in the 'task' of constructing the heritage of the cute (Barua, 2016, p. 730).

Beginning with a posthumanist evaluation of the classical Marxist theory of commodification and value exchanges, Haraway (2013) argues that Marx's study has resorted to human exceptionalism. Instead, Haraway shows a 'trans-species relation' (2013, p. 46) has been critical to contemporary society, where encounter value between species offers an additional dimension to both the use value and exchange value of a commodity, building on an array of relationships among the lively beings. Barua further clarifies the idea of encounter value, suggesting that the notion includes 'processes of growth, of working with materials, where nonhuman animals play an integral role' (2016, p. 729). Barua (2016) suggests non-human animals are performing tasks rather than labouring or working as their intentional human counterparts. The animals play an integral role in human–animal rhythms, producing intransitive value through eventful encounters. Barua (2020) extends the tasks of performing animals in his study of encountering giant pandas in captivity, specifying the 'affective labour' that panda bears perform through corporeal and somatic means.

In this chapter, I attempt to develop Barua's (2016, 2020) understanding of Haraway's encountering value on two points. First, the non-human's participation in encountering value is recognized through the tasks the animals perform. The task explicates a materialistic perspective of encountering value, highlighting the bodily coexistence between animals and humans, or the animal's 'liveliness' (Barua, 2020, p. 8). Second, I show that the affective labour of non-humans recognizes the contribution of immaterial emotions and affects that circulate around animal celebrities such as the giant panda, or 'affective atmospheres' in Barua's (2020, p. 13) term. The immaterial perspective, or the affective labour, shapes a human–animal relationship that becomes an economy itself. The perspective challenges the view that affects can be identified within a non-human but insists that the non-human is thus objectified because of its link to the affect.

The concept of lively capital helps to understand the tourist–panda relationship in the light that the pandas' cuteness can be identified as a currency circulating between the tourists and the animals. The cuteness indicates a particular force of affection, shaping the movements between the tourists and pandas. What becomes more interesting, however, is the presence or embodiment of 'the animal other' amid the circulation of affects. For Ahmed (2004, p. 124), the non-human other has a paradoxical presence that both experiences objectification while escaping the embodiment. For example, Ahmed explains that the emotion of fear 'is intensified by the impossibility of containment', rather than 'being contained in an object' (2004, p. 124). Ahmed's point is that, precisely because of its inability to be contained, fear escapes the other, returns to the subject and furthers its circulation. Likewise, the panda in affective economies has been objectified while escaping the effect of objectification. Cuteness encapsulates the tourist and the panda, but neither the tourists nor the panda can contain the affects and emotions of the cute fully, generating a surplus value that accumulates across history (Barua, 2020) and, in this study, building up a foundation for the heritage of the cute.

The Panda–Tourist Relationship Through the Cute: A Posthumanist Critique

The heritage of the cute is a result of the accumulating effect of affective economies that has dictated the relationship between tourists and pandas on a global level. The experience of the cute underscores a movement between the pandas and tourists that can be best characterized as 'an adoration of the commodity' (Ngai, 2005) – the tourist consumes the images of the panda. It means that pandas are not naturally and inherently cute, but that the cute has enabled a consumptive relationship between the panda and the tourists. The pandas are, instead of being cute, cutified animals who have been captured under the spotlight of consumerism.

On the surface, the consumption of the heritage of the cute and the conservation of the heritage of nature have been activities with opposite consequences. Traditionally, researchers (e.g. Higginbottom and Tribe, 2004; Ballantyne *et al.*, 2007; Ballantyne *et al.*, 2009) have searched for means to fill the gap between activities concerning consumption and conservation. One solution that has become prominent in both practice and research contexts is setting conservation and education as the goals of wildlife tourism (Larm *et al.*, 2018; Newsome, 2020; Macdonald and Wester, 2021). A few scholars (Belicia and Islam, 2018) have stressed the need to de-commodify wildlife in tourism. Researchers have also noted the possibility of building up alternative relationships between wildlife and the human population through the idea of cute. For example, in a study of the domestication of the Siberian silver fox, Dale *et al.* (2016), showing that the foxes are adapting to the idea of cuteness through a change in their appearances, argues that cuteness and aestheticization through the cute have initiated intra- and interspecies relationships. Dale *et al.* (2016) suggests the silver fox has demonstrated the ability of self-domestication, which increases the animal's friendliness towards people and contributes to the possibility of coexistence. In Dale *et al.*'s case, the silver fox has become an embodiment and containment of the cute that 'wants to engage with us, too' (2016, p. 51).

However, the panda's relationship with tourists has taken neither path. At the centre of the heritage of the cute, the panda and its images serve the interests of commodification and consumerism on a global level. Also, the panda's reluctance to breed in captivity (Diamond, 2002, p. 702) categorizes the animal as a non-domesticated species. While having kept its wildness intact, the panda journeys the farthest into the consumption and capitalization of the cute. As Sir Peter Scott (WWF, 2006), the designer of WWF's logo, stated, the charm and charisma of the giant panda have made the animal an ideal candidate for wildlife conservation. The consumption of cute pandas, from the very beginning, has been entangled with the conservation of wildlife. May, in her study on the power of the cute, suggests that the conventional views have only considered the innocent, tamed and helpless aspect of the cute objects, while the cute offers a dialogue between 'the familiar and unfamiliar, the unthreatening and the menacing, the present and the withdrawn, the visible and the invisible' (2019, p. 25). The undomesticated panda, a combination of the cute commodity and the untamed wild monster, embodies what Allison has observed in Japan's Pokemon monsters, namely, 'a play environment that is imaginary but also emotionally real' (2003, p. 389).

In this light, the conservation and consumption of the giant panda are inextricably linked and impossible to separate. The heritage of the cute provides a foundation on which conservation movements become a possibility, rather than the other way around. Affective economics surrounding the cute affords a cultural and imaginative space where the conservation and commodification of pandas can coexist through a mutually strengthening circulation. Because of its root in the nurturing of the vulnerable other, the roles of consumers and conservationists conflate in such a way that conservation has depended on cuteness for its further influence. The deep entanglement between conservation and commodification means that understanding the heritage of the cute can be the key to reconciling the roles of consumers and conservationists that tourists must play. The heritage of the cute opens a gateway towards a posthumanist tourist–panda relationship where the commodification and conservation of the pandas can be seen as two sides of the same coin. As the attraction of the cute has become inescapable for both tourists and pandas, symptomizing what May calls 'modernity's attraction to the innocence of becoming' (2019, p. 186), the cute provides a way to find meaning in modern life. What tourism researchers can focus on is the making of the cute heritage that can benefit both humans and pandas alike.

Method

In order to examine the giant panda as a heritage of cuteness in both urban and rural China, the author engaged in daily participant observation of the research topic. According to Musante and DeWalt, participant observation permits the researcher to 'take part in the daily activities, rituals, interactions, and events of a group of people as one of the means of learning the explicit and tacit aspects of their life routines and their culture' (2010, p. 1). As a qualitative method, participant observation provides the researcher with insight into the significance of a phenomenon. The heritage of the cute, which has had a significant impact on the urban and rural lives of the Chinese, is best understood in its naturalistic settings. The author, working at Chengdu Research Base of Giant Panda Breeding, kept field notes of her work, life, fieldwork and travels throughout the city. Photos, videos, travel brochures and pamphlets were gathered and organized into a collection that aided the author in understanding the heritage of the cute. For example, as a citizen of Chengdu, the author collected images of giant pandas around the city, bearing in mind the idea of seeing how Chengdu has incorporated the giant pandas and their representations into its city life. This idea was initially documented on the author's own mobile phone and emerged as a well-themed collection over time. As theoretical themes began to emerge, this collection was further grouped and examined. Using this collection of the heritage of the cute, I will demonstrate how the cute image of the panda has become integral to lives of the locals and tourists.

Consuming and Building the Heritage of the Cute in Urban China

'我们在熊猫屁股下面见 *Wǒmen zài xióngmāo pìgu xiàmiàn jiàn*', literarily translating into 'we meet under the panda buttocks' and a phrase specifying a must-see place in Chengdu downtown (Wei, 2020), refers to a wall-climbing giant panda sculpture on the top of Chengdu International Finance Square (Chengdu IFS) (Fig. 12.1). Created by the artist Lawrence Argent, the panda sculpture, named 'I am Here', was introduced to the Chengdu public on 14 January 2014, in the opening ceremony of IFS. The sculpture soon gained popularity among the public and became a highlight of the city. Identifying itself as the 'most international urban complex in Western China' and 'a landmark development' (IFS, 2022), the climbing panda merges with international brands such as Dior, UNIQLO, Prada, Basic House and Dolce & Gabbana. Despite its habitat in the deep forest of Sichuan, the panda sculpture speaks the language of fashion and holds the spirit of consumerism through its incarnation of the heritage of the cute.

In her study of images of China in American films, Greene observed the Kung Fu Panda balloon floating across New York during the 2010 Macy's Thanksgiving Day Parade, and noted that the panda figure has been 'freed from his origins in China' and 'joined the ranks of figures like Spider-Man and Shrek, Snoopy, and Dora the Explorer' (2014, p. 213). Greene's (2014) study of Kung Fu Panda is the transformation of wildlife in China into an American icon through filming and narratives. On the top of Chengdu IFS, the climbing panda sculpture claims a similar experience. Rather than serving as a link to its natural habitat

Fig. 12.1. Chengdu IFS climbing panda sculpture. (Photos by author.)

and the need for conservation, the climbing panda denotes the desires, passion and needs behind fashion brands, transforming people into consumers. According to Argent, the panda sculpture 'declares its presence and gives the art installation social meaning for thought, hence making ourselves think about our presence and reflecting on thought-provoking social significance' (2021, np). The panda sculpture links people to the making of a self and thus speaks the language of the consumers rather than the animal itself. The panda sculpture celebrates the heritage of the cute together with citizens whose social identity completes on their transformation into consumers.

The heritage of the cute enables an imaginative space where tourists can place themselves under the buttocks of the animal, while the place transforms people into consumers and urbanites in the city centre of Chengdu. The cuteness of the giant panda captures visitors through an intense emotional exchange that, more than presenting the animal's existence, returns a self-evaluation affirming the presence of the tourists themselves (Ahmed, 2004). Fig. 12.2 demonstrates a collection of 'pandas' in Chengdu city on a variety of buildings and public facilities. The cuteness of the animal contributes to its omnipresence in the city, stipulating emotional capital (Andrejevic, 2011) that continues to circulate between the image of the giant panda and people. In the city, the panda has not been able to contain its cuteness. Instead, the cuteness of the giant panda accumulates through the intense emotional exchanges between the tourists and the animal, building a heritage of cuteness that has become a significant asset of the city.

In urban Chengdu, the heritage of the cute has thematized the presence of the giant pandas, uprooting the animal from its natural habitat and making it a symbol of its own. The cuteness of the panda transforms citizens and tourists into consumers united under a shared identity. The cuteness of the animal has escaped from its material body but is a symbol of modern life on its own. The panda and its presence in rural China, however, have been a different story.

Meeting the Conservation–Consumption Gap in Rural China

Researchers have consistently noted a prominent gap between rural and urban China that has been sufficient to form two different societies in the nation (Guang *et al.*, 2010). The rural becomes a source of attraction for the urban (Wang *et al.*, 2013) for its provision of natural resources, rurality, landscapes and biodiversity that have become more attractive with the growth of urbanization in China. In this study, the rural region observed is Longcan Valley, a remote village 193 km away from Chengdu at the southern entrance of the Giant Panda National Park in Yingjing County. With 80% forest coverage, Longcanggou Valley is famous

Fig. 12.2. Pandas in the city of Chengdu. (Photos by author.)

for its rich natural resources and biodiversity. Contributing to the protection of the wildlife population and the restoration of its habitat, Yingjing County intends to construct a green barrier to safeguard the natural environment. As coal mines and hydropower stations were removed from Longcanggou Valley in the early 2000s, locals sought alternative means of establishing a livelihood. Three villages in Yingjing County have transformed themselves from an agrarian society into a tourism hotspot for visitors from Chengdu and nearby cities. The first '*nongjiales*' (the bed and breakfast homestay business in rural China) were established in 2008, and by 2021 the local residents in three villages owned 64 *nongjiales* in total.

While the story of the rural region's transformation has been remarkable in the past two decades, Fazhan Village took the initiative to declare its relevance with the giant panda in order to better promote the place to urban visitors. In 2020, the village committee coined the term 'Panda Homestay Village' and constructed screen walls (Fig. 12.3) with the name emblazoned on them to inform tourists on their arrival. Paintings and sculptures depicting pandas were added to *nongjiales* with funding from the village committee. Fazhan Village developed a connection to the giant panda in order to attract more tourists, despite the fact that encountering giant pandas in Fazhan Village, like other locations in the national park, is restricted to stories told by local elders. The immediate link that Fazhan Village saw between its homestay business and the giant panda is the heritage of the cute rather than conservation efforts designed to reduce habitat fragmentation and boost biodiversity.

On the rural roads, panda figures (Fig. 12.4) can be sighted. On the surface, Fazhan Village's celebration of panda images is similar to Chengdu's passion for the animal. It can be argued that a rural place like Fazhan Village is aware of the heritage of the cute and, by incorporating panda images and names, tries to attract more urban visitors. Despite the low likelihood of encountering wild pandas in

Fig. 12.3. Panda Homestay Village screen walls at the entrance of Fazhan Village. (Photos by author.)

Fig. 12.4. Pandas on the roadside in Fazhan Village. (Photos by author.)

the village, the village committee believed that branding the village with panda images would provide an emotional attraction and affinity with visitors that was worth investing in. The heritage of the cute attempts to integrate a rural village into the effects of consumption, which contribute to the influx of tourists into rural areas. The affective economies of the cute pandas create an atmosphere of consumption that Fazhan Village attempts to materialize through the tourism business.

Rather than climbing on the top of a financial centre, in Fazhan Village, the giant panda sculptures were placed amid bushes where nature and human society crossed on the rural roads. The pandas look at passengers on the road from a distance that, having had its cultural and urban capital abolished, tells the remaining gap between consumption and conservation. Disoriented and displaced, the panda sculpture in the bush found itself both an exclusion from the human world and the natural habitat to which it belongs – the pandas greeting the urban visitors have been fake figures, having neither the proper form of the wildlife nor its fully capitalized counterpart on Chengdu IFS. Haraway believes that lively capital is a product of tightly interwoven naturecultures. Chengdu and Fazhan Village have both attempted to mobilize the giant panda's lively capital. While Chengdu IFS has successfully made an impact through the cute image of the giant panda, Fazhan Village, it seems, works on the still fragmented fringe of nature and culture.

Conclusion

The giant panda and its ecological value have received great attention since the animal's discovery in Baoxing, Sichuan in 1869. However, the recognition of giant pandas in human society has not been a simple story of conservation efforts and scientific knowledge generation. This chapter explores the origin and construction of the heritage of the cute that affective economies allow to circulate and accumulate through the form of the lively capital of the giant panda and its importance in contemporary urban and rural China. The affective economies underscoring the heritage of the cute provide a mechanism of capitalization through which affects towards cute pandas are produced and reproduced. The heritage of the cute indicates a continuous process of accumulation as affects are exchanged and communicated between pandas and tourists, deepening an us–them division. At the same time, the cuteness of the giant pandas mobilizes people, transforming them into consumers who share a mutual identity and recognition, while pandas become animals embodying people's passion and desire for consumption. The heritage of the cute marks the accumulation and appreciation of lively capital in affective economies, strengthening both the reproduction of lively capital and the shared bond between consumers.

The heritage of the cute claims the city image of Chengdu, celebrating consumerism and a social identity connecting the city to the rest of the world. The climbing panda on Chengdu IFS materializes the heritage of the cute, integrating with the construction of the identity of consumer selves. Surrounded by fashion brands, the climbing panda makes a break with its natural habitat in the forest and conservation efforts, but its towering influence over the city of Chengdu urges a critical examination of the relationship between conservation and consumption. This study manifests that the image of the cute can be fundamental to both consumers and conservationists because affective economics also organizes the order and activities of panda conservation. Despite its negligence in current research, the heritage of cuteness serves as an appeal at the heart for both consuming and conserving the pandas. A posthumanist tourist–panda relationship, where commodification and conservation of the species are intricately intertwined, will find its path building on the heritage of the cute through the intense exchanges within affective economies. The cuteness becomes a heritage we may investigate when it shows that there is a way to bring the two species together, and in the case of the giant panda, this encounter provides the endangered species with a crucial opportunity for survival. A better understanding of the structure and organization of the heritage of the cute will provide more opportunities where species can meet. It is out of this commodified, capitalized and often unequal interspecies encounter that conservation efforts and programmes can grow.

However, I believe it is the panda in the rural Fazhan Village that has a more truthful story to tell between the current gap in what we consider to be the consumption and conservation of wildlife. Having no finance centre to climb over, pandas in rural areas are left to a wilderness to which they, as a heritage of the cute, fail to belong immediately. While wild pandas today still experience habitat fragmentation (Yang *et al.*, 2020; Connor *et al.*, 2021), deteriorating ecosystems (Li *et al.*, 2020) and biodiversity loss (Li, 2020), their cute counterparts seem to be indifferent to these stories. Most importantly, because the heritage of the cute originates in the affects and emotions of consumers, it draws a boundary between them and the wild pandas based on their interests and affects, shaping bodies that meet the desires and needs of people rather than animals. While the wild pandas remain silent and unseen behind the dense bushes in the deep forest in Sichuan, tourists and consumers are left with roadside cute panda figures looking out at people themselves.

References

Ahmed, S. (2004) Affective economies. *Social Text* 22(2), 117–139. DOI: 10.1215/01642472-22-2_79-117.

Allison, A. (2003) Portable monsters and commodity cuteness: Pokémon as Japan's new global power. *Postcolonial Studies* 6(3), 381–395.

Andrejevic, M. (2011) The work that affective economics does. *Cultural Studies* 25(4–5), 604–620. DOI: 10.1080/09502386.2011.600551.

Argent, L. (2021) I Am Here. Available at: https://lawrenceargent.com/public-art-projects/i-am-here/#:~:text=With%20the%20grand%20opening%20of,a%20height%20of%2015%20meters (accessed 5 May 2022).

Ballantyne, R., Packer, J., Hughes, K. and Dierking, L. (2007) Conservation learning in wildlife tourism settings: lessons from research in zoos and aquariums. *Environmental Education Research* 13(3), 367–383. DOI: 10.1080/13504620701430604.

Ballantyne, R., Packer, J. and Hughes, K. (2009) Tourists' support for conservation messages and sustainable management practices in wildlife tourism experiences. *Tourism Management* 30(5), 658–664. DOI: 10.1016/j.tourman.2008.11.003.

Barua, M. (2016) Lively commodities and encounter value. *Environment and Planning D: Society and Space* 34(4), 725–744. DOI: 10.1177/0263775815626420.

Barua, M. (2020) Affective economies, pandas, and the atmospheric politics of lively capital. *Transactions of the Institute of British Geographers* 45(3), 678–692. DOI: 10.1111/tran.12361.

Belicia, T.X.Y. and Islam, M.S. (2018) Towards a decommodified wildlife tourism: why market environmentalism is not enough for conservation. *Societies* 8(3), 59. DOI: 10.3390/soc8030059.

Burns, G.L. (2015) Animals as tourism objects: ethically refocusing relationships between tourists and wildlife. In: Markwell, K. (ed.) *Animals and Tourism: Understanding Diverse Relationships*. Channel View Publications, Bristol, UK, pp. 44–59. DOI: 10.21832/9781845415051.

Carr, N. and Broom, D.M. (2018) *Tourism and Animal Welfare*. CABI, Wallingford, UK. DOI: 10.1079/9781786391858.0000.

Connor, T., Qiao, M., Scribner, K., Zhang, J., Hull, V. *et al*. (2021) Complex effects of habitat amount and fragmentation on functional connectivity and inbreeding in a giant panda population. *Conservation Biology* 36(2), e13828. DOI: 10.1111/cobi.13828.

Dale, J.P. (2016) The appeal of the cute object. In: Dale, J.P., Goggin, J., Leyda, J., McIntyre, A.P. and Negra, D. (eds) *The Aesthetics and Affects of Cuteness*. Routledge, New York, pp. 35–55. DOI: 10.4324/9781315658520.

Dale, J.P., Goggin, J., Leyda, J., McIntyre, A.P. and Negra, D. (2016) The aesthetics and affects of cuteness. In: Dale, J.P., Goggin, J., Leyda, J., McIntyre, A.P. and Negra, D. (eds) *The Aesthetics and Affects of Cuteness*. Routledge, London, pp. 11–44. DOI: 10.4324/9781315658520.

Diamond, J. (2002) Evolution, consequences and future of plant and animal domestication. *Nature* 418(6898), 700–707. DOI: 10.1038/nature01019.

Fennell, D.A. (2013) Tourism and animal welfare. *Tourism Recreation Research* 38(3), 325–340. DOI: 10.1080/02508281.2013.11081757.

Genosko, G. (2005) Natures and cultures of cuteness. *InVisible Culture* 9. DOI: 10.47761/494a02f6.b4abc93a.

Greene, N. (2014) *From Fu Manchu to Kung Fu Panda: Images of China in American Film*. University of Hawaii Press, Hawaii. DOI: 10.21313/hawaii/9780824838355.001.0001.

Guang, L., Adams, J., Gaetano, A., Whyte, M.K., Gustafsson, B. *et al*. (2010) *One Country, Two Societies: Rural-urban Inequality in Contemporary China*, Vol. 16. Harvard University Press, New York.

Guo, Y. (2021) 'Cute, but get up and work!': the biophilia hypothesis in tourists' linguistic interactions with pandas. In: Rickly, J. and Kline, C. (eds) *Exploring Non-Human Work in Tourism: From*

Beasts of Burden to Animal Ambassadors. De Gruyter, Oldenbourg, Germany, pp. 225–241. DOI: 10.1515/9783110664058.

Haraway, D.J. (2013) *When Species Meet*, Vol. 3. University of Minnesota Press, New York.

Harris, D. (2000) *Cute, Quaint, Hungry, and Romantic: The Aesthetics of Consumerism*, Vol. 10. Basic Books, New York.

Higginbottom, K. and Tribe, A. (2004) Contributions of wildlife tourism to conservation. In: Higginbottom, K. (ed.) *Wildlife Tourism: Impacts, Management and Planning*. Common Ground Publishing, Melbourne, Australia, pp. 99–123.

IFS (2022) About us. Available at: https://www.cdifs.cn/en/about-ifs (accessed 2 May 2022).

Kline, C. (2018) Abstracting animals through tourism. In: Kline, C. (ed.) *Tourism Experiences and Animal Consumption: Contested Values, Morality and Ethics*. Routledge, London, pp. 208–217. DOI: 10.4324/9781315265186.

Larm, M., Elmhagen, B., Granquist, S.M., Brundin, E. and Angerbjörn, A. (2018) The role of wildlife tourism in conservation of endangered species: implications of safari tourism for conservation of the Arctic fox in Sweden. *Human Dimensions of Wildlife* 23(3), 257–272. DOI: 10.1080/10871209.2017.1414336.

Li, B.V. (2020) Conservation: guarding panda land. *Current Biology* 30(7), R302–R304. DOI: 10.1016/j.cub.2020.02.089.

Li, S., McShea, W.J., Wang, D., Gu, X., Zhang, X. *et al.* (2020) Retreat of large carnivores across the giant panda distribution range. *Nature Ecology & Evolution* 4(10), 1327–1331. DOI: 10.1038/s41559-020-1260-0.

Lukacs, G. (2015) The labor of cute: net idols, cute culture, and the digital economy in contemporary Japan. *Positions* 23(3), 487–513. DOI: 10.1215/10679847-3125863.

Macdonald, C. and Wester, J. (2021) Public understanding of wildlife tourism: defining terms, harms, and benefits. *Journal of Ecotourism* 20(2), 198–209. DOI: 10.1080/14724049.2020.1817930.

May, S. (2019) *The Power of Cute*. Princeton University Press, New Jersey. DOI: 10.1515/9780691185712.

Musante, K. and DeWalt, B.R. (2010) *Participant Observation: A Guide for Fieldworkers*. Rowman Altamira, Lanham, Maryland.

Newsome, D. (2020) The collapse of tourism and its impact on wildlife tourism destinations. *Journal of Tourism Futures* 7(3), 295–302. DOI: 10.1108/JTF-04-2020-0053.

Ngai, S. (2005) The cuteness of the avant-garde. *Critical Inquiry* 31(4), 811–847. DOI: 10.1086/444516.

Nicholls, H. (2012) *The Way of the Panda: The Curious History of China's Political Animal*. Pegasus Books, New York.

Rickly, J.M. and Kline, C. (2021) Working for the (hu)man in the tourism industry. In: Rickly, J. and Kline, C. (eds) *Exploring Non-Human Work in Tourism: From Beasts of Burden to Animal Ambassadors*. De Gruyter, Berlin, pp. 1–20. DOI: 10.1515/9783110664058.

Songster, E. (2018) *Panda Nation: The Construction and Conservation of China's Modern Icon*. Oxford University Press, Oxford, UK.

Wang, L.-E., Cheng, S.-K., Zhong, L.-S., Mu, S.-I, Dhruba, B.G.C. *et al.* (2013) Rural tourism development in China: principles, models and the future. *Journal of Mountain Science* 10(1), 116–129. DOI: 10.1007/s11629-013-2501-3.

Wei, F., Costanza, R., Dai, Q., Stoeckl, N., Gu, X. *et al.* (2018) The value of ecosystem services from giant panda reserves. *Current Biology* 28(13), 2174–2180. DOI: 10.1016/j.cub.2018.05.046.

Wei, W. (2020) The construction of Chengdu city image in new media environment: take tik tok as an example. In: *6th International Conference on Humanities and Social Science Research (ICHSSR 2020)*, Hangzhou, China. DOI: 10.2991/assehr.k.200428.102.

Winter, C. (2020) A review of research into animal ethics in tourism: launching the annals of tourism research curated collection on animal ethics in tourism. *Annals of Tourism Research* 84, 102989. DOI: 10.1016/j.annals.2020.102989.

WWF (2006) Panda sanctuary in China added to world heritage list. Available at: https://wwf.panda.org/wwf_news/?76460/Panda-sanctuary-in-China-added-to-World-Heritage-list (accessed 5 May 2022).

Yang, B., Qin, S., Xu, W., Busch, J., Yang, X., *et al.* (2020) Gap analysis of giant panda conservation as an example for planning China's national park system. *Current Biology* 30(7), 1287–1291. DOI: 10.1016/j.cub.2020.01.069.

Yano, C.R. (2013) *Pink Globalization*. Duke University Press, Durham, North Carolina. DOI: 10.1515/9780822395881.

13 Contested Boundaries: Towards a Posthumanist Political Ecology Approach to Recognize Human–Deer Relationship as an Intangible Cultural Heritage

Rie Usui[1]* and Carolin Funck[2]

[1]*Ritsumeikan Asia Pacific University, Oita, Japan;* [2]*Hiroshima University, Higashihiroshima, Japan*

Abstract

This chapter explores human and non-human animal territorial conflicts at a cultural tourism destination through the lens of posthumanist political ecology. It analyses how space is contested by attending to interspecies interactions on Japan's Miyajima Island, specifically, human–deer interactions. The study shows that humans frame deer into the 'wild' space, removing the deer from the place where they belong, although this imagined boundary is constantly crossed by the deer and humans. This evidence indicates that the deer are active agents and the existing landscapes of Miyajima Island have, to a certain extent, been influenced through constant interactions with humans and the deer. Recognizing close human–deer relationship as one essential component of heritage in Miyajima would help us coexist with the deer.

Introduction

Cultural tourism – the context of the present research – refers to tourists' search for both tangible and intangible cultural products, such as architecture, historical and cultural heritage, music, arts and culinary heritage (Richards, 2018). Due largely to the connotation of culture, non-human animals may be relegated to the 'background' (Catlin *et al.*, 2013) at many cultural tourism destinations, even though they play a significant role in making tourism landscapes. As some tourism scholars have emphasized recently, tourism is more-than-human practices (Dashper, 2020; Danby *et al.*, 2019; Thomsen *et al.*, 2021). Yet non-human animals are often simply exploited and, in some cases, killed, and are excluded from moral considerations. In fact, the approach of much of the tourism and leisure research has been anthropocentric (Danby *et al.*, 2019).

Japanese tourism research follows a similar trend, especially in the context of cultural tourism. Japan has a rich cultural heritage; a total of 20 sites in the country have been designated United Nations Educational, Scientific and Cultural Organization (UNESCO) World Cultural Heritage Sites (Agency for Cultural Affairs, Government of Japan, 2022), placing Japan in the top ten of the world ranking (UNESCO World Heritage Convention, 2022).

*Corresponding author: rieusui@apu.ac.jp

 Tourism, Heritage and Commodification of Non-human Animals (eds A. López López *et al.*)
DOI: 10.1079/9781800623309.0013

Fig. 13.1. Itsukushima Shrine and the red *torii* gate. (Photo by Rie Usui.)

Among them, Miyajima Island is well known for the Itsukushima Shrine and the red *torii* gate standing in the sea (Fig. 13.1). Furthermore, the free-ranging Japanese deer inhabiting the island provide a unique experience as they are tame and share a space with humans. However, an increase in the deer population and Miyajima's development as an international tourism destination prompted a rethinking of humans' relationship with the deer, leading to an agreement among officials to decrease the deer population and separate the human and deer living spaces (Usui and Funck, 2017, 2018). For the last ten years, the management officials have been trying to achieve this goal by eradicating human food handouts, but this strategy remains ineffective.

The above case exemplifies a human and non-human animal conflict, typically called a human–wildlife conflict, referring to adverse effects (e.g. threats to life, well-being and safety of both actors) resulting from interactions between humans and non-human animals (Conover, 2002; Treves and Karanth, 2003; Woodroffe *et al*., 2005). How to mitigate interspecies conflicts and coexist with non-human animals and nature in general has gained heightened interest, as is evident in the growing number of publications on the topic (Nyhus, 2016). The idea of separating the spaces occupied by humans and non-human others has spread worldwide and become deeply rooted in wildlife management despite the wide recognition that humans are a part of the environment (Iketani, 2008).

In a case such as that of Miyajima, where relationships between humans and non-human animals have been religiously influenced (Usui and Funck, 2018), a dualistic approach to wildlife management may not be compatible with the context they are in. Several factors presumably contributed to the failure of Miyajima's management policy, but whether separating the human and deer spaces is a feasible approach in the first place is a critical question given the history of close human–deer interaction that has existed for hundreds of years. The landscape of Miyajima, which has World Heritage value, has been co-created, to a certain extent, as a result of human–deer interactions. It is a landscape shared by humans and non-human others, including Miyajima's deer.

Place and space are central to humans' relationship with non-human animals. The human view of non-human animals is reflected in the classification of non-human animals, which places them in abstract places (Philo and Wilbert, 2000) and consequently excludes them from moral considerations (Braidotti, 2013; Kline, 2022). In this chapter, we argue that Miyajima's human–deer conflict is a matter of humans' conceptual construction of space. Particularly relevant to the case is the concept of wild that is associated with Miyajima deer. Conceptually and discursively, wilding the deer dismisses 'who' they truly are and deflects from the root cause of human–deer conflicts. Solving this issue requires attention to a different way of thinking and understanding the world. To this end, we aim to explain the implications for ethical management decision-making that balances the needs and interests of both humans and deer.

The present study was informed by posthumanist political ecology (Sundberg, 2011), which provided the theoretical grounding and analytical framework. Political ecology is an interdisciplinary field focused on human–environmental interactions and specifically examining social relations and subsequent power structures as a product of interactions between society and nature (Escobar, 1996). A fundamental stance in political ecology is that social–environmental interactions cannot be fully comprehended without considering political influence (Bryant, 1998). It is highly relevant to tourism (Nepal *et al.*, 2018) because tourism impacts the environment at a global level (Stonich, 1998). Political ecology is useful for examining the relationship between tourism and global-level environmental governance and management (Stonich, 1998; Cole, 2012).

Besides its holistic perspective that recognizes the inter-relationships between society and environment, political ecology can provide a useful analytical tool to examine the ethical treatment of non-human animals in the context of tourism for three reasons. First, it takes the historical context into consideration when examining environmental issues (Mostafanezhad *et al.*, 2016). Humans and non-human animals have been interacting with each other for thousands of years. Thus, the present form of interaction is the product of a historical process, making historical contextualization essential to understand how a certain relationship between humans and non-human animals was formed.

Second, political ecology decodes various usages of words and phrases, such as nature, conservation, wilderness and biodiversity, to understand how certain activities are performed (Douglas, 2014; Mostafanezhad *et al.*, 2016). For instance, Grimm (2013) explored how people involved in volunteer conservation tourism in Ecuador contextualize the word 'conservation' and found that different stakeholders had different understandings of it, which would then affect the outcome of the conservation project. Through discourse, certain practices and ideas become the norm (Barnes and Duncan, 1992, cited in Bryant, 2000). Discourse generated through language in the form of text or conversation (Neumann, 2004) can be examined to elucidate how non-humans are conceptually framed within a political system. However, the discursive aspect of society's interactions with the environment has received little research attention (Bryant, 2000).

Finally, by attending to stakeholders at various levels, political ecology implies that power structures are built and the dominant discourse governs environmental practices (Douglas, 2014; Mostafanezhad *et al.*, 2016). For instance, in examining different levels of 'the monkey problem' in West India, Dore (2018) found that nature conservation was idealized at a global level, which problematized human–monkey interactions at a local level. The study concluded that multi-scalar politics exist, which are attributable to the generation of stakeholder conflicts.

Although political ecology provides a contextual approach to understand interactions between the environment and society, and between various stakeholders within the society, non-human animals are often considered merely passive elements in this framework (Hobson, 2007). Tsing (2005) insisted that tourism scholars who employ political ecology should be more sensitive to non-human animals who play an essential part in tourism. Similarly, Witter (2013) criticized the lack of discursive political ecology research on exhibiting moral concern in conservation discourses.

To address the above shortcomings, Sundberg (2011) incorporated posthumanism into political ecology. Posthumanism, which emerged in the late 20th century, rejects humans' superiority over non-human animals and challenges basic ontological categories (Cohen, 2019). Viewing humans as one of many living beings in the world (Gaard, 2017), posthumanists engage with non-human others and emphasize non-human animals, rather than exclusively focusing on humans (Lloro-Bidart, 2017). Unlike ecofeminism or, more specifically, vegan eco/feminism, posthumanism does not require paying attention to intra-human inequities or gender (Gaard, 2017). Posthumanism is further characterized by its insistence that humans show empathy towards non-human animals (Castricano, 2008) and its emphasis on non-human animal justice, which distinguishes it from ecofeminism (Soper, 2012; Tomassini *et al.*, 2022).

By combining political ecology with posthumanism, Sundberg asserted that 'a posthumanist political ecology refuses to treat nonhuman nature as the thing over which humans struggle and instead builds on and enacts a relational approach in which all bodies are participants in constituting the world' (2011, p. 322). She added:

> An approach promises to animate and enliven the study of boundary making by attending to (1) the properties and capacities of the many embodied beings that bring the world into being, (2) everyday practices and doings-in-relation, and (3) specific sites of politics-in-action. (2011, p. 322)

In this study, we employ a posthumanist political ecology (Sundberg, 2011) to explore the human–deer conflict and relationship on Miyajima Island with a particular focus on the physical and abstract place and space that emerge with the term wildlife. Our aim is to unravel the assumption underlying wildlife in people's relationship with deer and to suggest an alternative form of coexistence between humans and deer that justifies the use of a multispecies landscape.

Methodology

A mixed-methods case study combining quantitative and qualitative instruments was used as it enriches the understanding of the complex relationships between various actors entangled in the political-ecological network (Turner, 2015). This study is based on fieldwork conducted on Miyajima Island between 2014 and 2018. During this four-year period, behavioural observations of interactions between tourists and deer were performed for 16 days and a questionnaire about tourists' perceptions of deer was administered (276 Japanese and 256 international tourists). In addition, interviews were conducted with various stakeholders, including retired workers, non-profit organization (NPO) workers, a restaurant owner's family, a local retired couple and 41 tourists. An ethnographic approach was also taken to observe and record the daily lives of human and deer residents. Additionally, the first author participated in the deer population census in 2017 and 2018 and attended a Miyajima Deer Management meeting in March 2018. Literature and documents were also collected, which provided basic information related to the study and materials that also served as data sources. Items such as official administrative documents, archival records, personal documents, pamphlets, maps, tourist information booklets and newspapers were also collected. Moreover, deer management plans for the Miyajima deer were retrieved from Hatsukaichi City's official website.

Case Study: Miyajima Island

Miyajima Island as a cultural tourism destination

Miyajima Island is located in Hiroshima Prefecture in south-western Japan (Fig. 13.2). Miyajima Island has been a setting for worship since ancient times (Miyajima Tourism Association, 2017), influenced by a religious tradition rooted in Shintoism. Traditionally, the Japanese considered nature, whether organic or inorganic (e.g. wind, thunder, sun, mountains, rivers, trees, rocks, animals, ancestral spirits), as an entity with spiritual power (Ono, 1962).

People considered the island itself to be a sacred place where gods reside and erected their

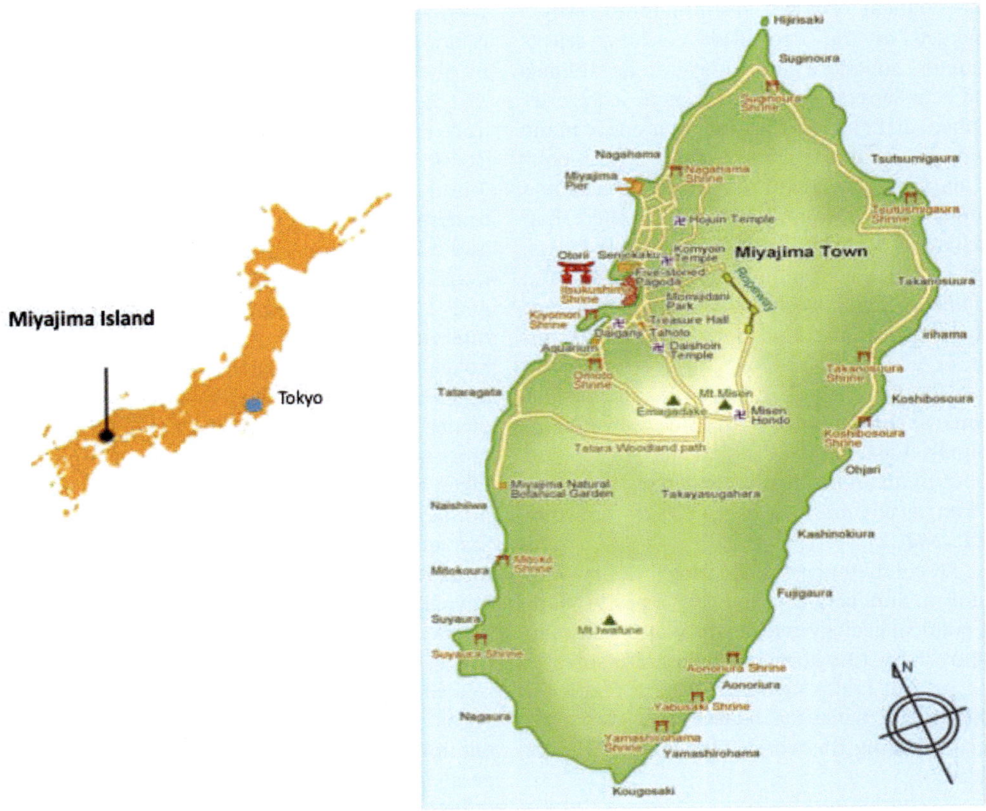

Fig. 13.2. The location of Miyajima Island in Japan and a map of the island. (Used with permission from Miyajima Tourism Association, 2017.)

first shrine as early as the 6th century (UNESCO, 2022). Thus, Miyajima Island remained protected for centuries, and this worldview has created a unique cultural landscape including, but not limited to, Itsukushima Shinto Shrine and the red *torii* gate built in the sea. In 1996, Itsukushima Shrine and the forest area surrounding it were registered as a World Cultural Heritage Site, and in 2012, a natural coastal wetland of the island was designated a Ramsar site (Ramsar, 2012). The tourism district (approx. 4.3 km^2) is located on the northern part of the island and the rest of the landscape is a forested area. Before the COVID-19 pandemic, as many as four million domestic and international tourists would visit the island annually (Hatsukaichi City, 2023).

History of human–deer interaction

Today's close association between humans and deer dates as far back as the 17th century. The deer inhabited Miyajima Island long before humans' arrival on the island (Funatsuki, 2011). Miyajima art and travel records depict deer resting with humans near shrines and in the streets, being fed by humans and strolling the streets. When Japan underwent social-political restoration from the mid-1800s to the early 1900s, the deer population declined. After the restoration period, a movement to protect the deer allowed the population to rebound. These events coincided with the first visits to Miyajima by foreigners, who recorded the deer's activities through their eyes. For example, Herbert George

Ponting, a British photographer who visited Miyajima Island in the early 1900s, wrote in his travel journal that deer came down from the mountains and walked along the shore, licking salt off the rocks and eating crackers visitors gave them (Ponting, 2005).

After the Second World War, the number of deer on Miyajima Island declined sharply, and in the 1960s, locals responded by placing the remaining deer in protective enclosures and raising deer to rebuild the population. In the mid-1960s, town officials announced a plan to release deer in the town. Their decision was motivated by the hope to recreate an old Miyajima townscape characterized by deer. The strolling deer soon became popular and were called 'cute' and 'adorable' (Chugoku Shinbun, 1968).

Reports on deer indicate that the human–deer relationship was somewhat positive up until the mid-1960s. The deer themselves and deer-related activities were referred to as a specialty of Miyajima. Within ten years, however, the deer population had grown so much that they became 'overabundant' (from the human perspective) and began creating conflicts with humans, including damage to gardens and plant displays, traffic accidents involving deer, disease transmission, injuries to locals and tourists, and damage to mountain vegetation and the ecosystem (Hatsukaichi City, 2014b). According to a population census, approximately 350 deer lived on the island – a density greatly exceeding the sustainable deer population capacity (Hatsukaichi City, 2014b). Despite the proliferation of deer, the Miyajima town officials wished to increase the deer number further for tourism. The residents, however, struggled with the situation because free-roaming deer would eat their plants and even laundry (Hatsukaichi City, 2014b). In 1978, former town officials organized a meeting to discuss a solution.

Management of human–deer interactions

Deer management on Miyajima Island has proceeded by trial and error. Table 13.1 shows a timeline of events regarding deer management on the island. During the first official town meeting concerning deer management in 1978,

Table 13.1. Timeline of deer management actions in Miyajima. (From: Hatsukaichi City, 2014b)

Year	Deer management action (Other events)
1965	(Deer population increased)
1972	Cutting male deer antler began to prevent injuries
1978	Miyajima Town meeting discussed the deer problem Population census was conducted and 350 deer were estimated Feeding station was set up in the mountains
1984	Male deer were confined to an enclosure
1987	Feeding station was removed
1998	Miyajima Town Deer Management Committee was assembled by Miyajima Town
2005	(Miyajima Town merged into Hatsukaichi City)
2006	Ministry of the Environment officials reported the negative impacts of the deer population increase on vegetation
2007	Hiroshima Prefecture conducted a deer population census on Miyajima Island and issued a report addressing the policy to ban feeding
2008	Hatsukaichi City Miyajima Region Deer Management Committee was assembled by Hatsukaichi City A survey was administered to local residents Miyajima Region Deer Protection Management Guideline was approved
2009	Miyajima Region Deer Protection Management Plan was proposed
2014	The Second Miyajima Region Deer Protection Management Plan was proposed

the officials agreed to introduce a feeding station in the forest to lure the deer to the adjacent mountains. This tactic was employed until 1987, but it failed to control the deer population (Hatsukaichi City, 2014b). In 1998, officials sought to reduce the number of deer in the tourism district by banning feeding. However, the prohibition was largely unsuccessful, and the practice of feeding continued to burden the local management authorities.

After the merger of Miyajima Town with Hatsukaichi City in 2005, Hatsukaichi City officials gained control over deer management and explicitly labelled the deer on Miyajima Island wild animals. In 2007, a campaign intended to discourage tourists from feeding the deer was initiated, but a street vendor continued to sell deer crackers to tourists until their sale was banned in 2008. That same year, in March, the Miyajima Region Deer Protection Management Committee, consisting of Hatsukaichi City officials, local representatives and scholars, was formed (Hatsukaichi City, 2014b). Since 2009, the Miyajima Region Deer Protection Management Plan has been in force.

The management officials have been working to reduce the total deer population for over a decade now (as of 2023). City officials' goal was to dehabituate the deer and encourage them to return to the mountains to reduce their effects on the tourism district. The prohibition on feeding deer was implemented as a means to achieve this goal (Hatsukaichi City, 2014b). While one reason for the feeding ban was to protect the deer from eating unhealthy food, the ban has not been enforced, rendering it ineffective. Moreover, as the officials continue to allow street vendors to sell food to tourists, such as French fries, curry bread and grilled oysters, which attracts deer, their consideration for deer welfare remains questionable. In addition, the implementation of sterilization has been contemplated, albeit not publicly.

Deer have been removed from Miyajima's tourism map as well. The maps below were issued in May 2006 and May 2012, respectively (Fig. 13.3). In comparing these maps, it becomes clear that the 2012 map shows a very different image from the 2006 map. On the basis of the deer management history briefly reviewed in this section, Miyajima deer are viewed as a nuisance rather than an element of heritage.

Multiple stakeholders' views on deer: residents, tourists and activists

There seem to be barriers to successfully accomplishing Miyajima's deer management plan, as multiple layers of conflicts and challenges exist among the local management authorities, residents, tourists and opposition groups.

Residents

Residents of Miyajima encounter deer on a regular basis, and their attitudes towards the deer and the deer management policy are heterogeneous. Some residents have formed an attachment to particular deer, and others are indifferent to the deer. A few residents dislike the deer's presence. One of the respondents said:

> On one hand, there is a group that wants the deer to go away from the residential areas, because if the deer cause trouble with their customers, they do not want to deal with it. On the other hand, there is a group of people who have lived with the deer and consider such treatment cruel. There are largely two groups. The latter group has asked Hatsukaichi City to reconsider the management policy multiple times, but the town officials do not want deer ... Currently, multiple opinions are entangled, and the situation is such a mess. Residents attend meetings held by the management authority as observers. However, the observers are not allowed to say anything. (Respondent A, 2018)

As respondents explained, two polarized opinions exist, and the observation of residents' interactions with the deer supported this. During the fieldwork in Miyajima, we often observed residents feeding deer in the residential areas. One respondent, who has run a business in Miyajima for decades, stated:

> I am often told not to feed the deer. People probably have eyes on me because I always feed the deer. We get lots of vegetables that we need to get rid of. I keep extra cabbage and lettuce for

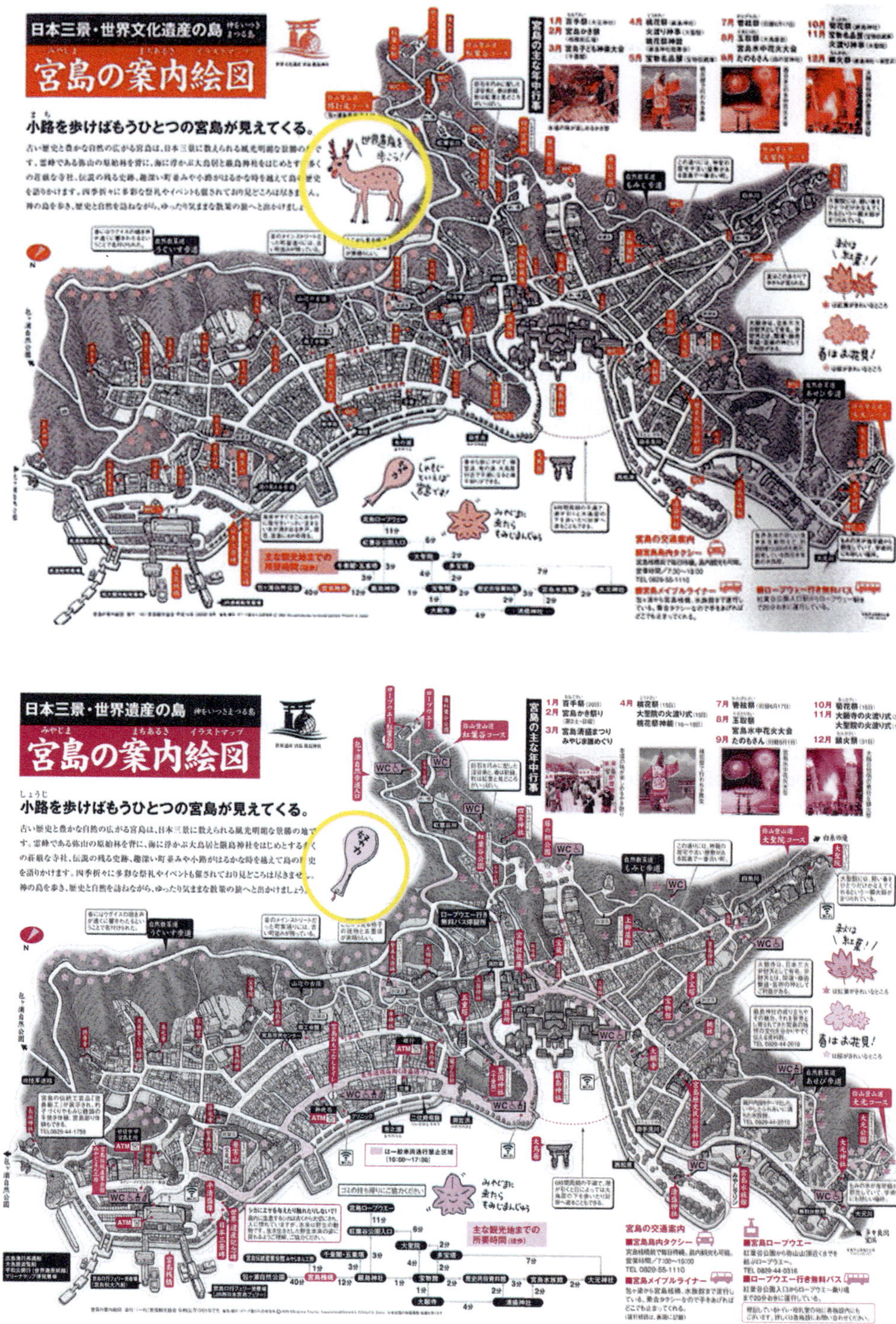

Fig. 13.3. Miyajima maps of 2006 (top) and 2012 (bottom).

> the deer.... I have been giving food to the deer for years now. ... My friend who owns a restaurant in Miyajima said she feeds the deer secretly too. (Respondent B, 2018)

Despite residents' remarks that deer are cute, the deer can be troublesome. One female resident reported that the deer in Miyajima annoy her sometimes:

> When tourists see the locals chasing after deer with brooms and sticks, they say 'Poor deer', but if they live here for three days, they will understand why we do that. The deer damage is significant. (Respondent C, 2014)

She found the deer's behaviour troublesome not because she was adversely affected but because behaviours such as eating flyers could do damage to the deer themselves.

Tourists

Tourists expressed mixed views regarding the status of the deer (Table 13.2). Fifty-four percent of Japanese and 50.8% of international tourists consider the deer semi-wild. Significantly more Japanese tourists (n=61, 22.1%) consider them wild than do international tourists (n=34, 13.5%).

The results indicated that the status of Miyajima deer is ambiguous to tourists. Most tourists (more than 50% of both Japanese and international tourists) consider the deer 'semi-wild'. This is likely because the deer do not belong to a specific owner (i.e. a feature of wild non-human animals), yet they are habituated to humans (i.e. a feature of domesticated non-human animals). What is noticeable in the results is that significantly more Japanese consider the deer 'wild', while significantly more international tourists consider the deer 'not wild'. The concept of wilderness likely varies by country, cultural background and personal experience.

Table 13.2. Do you consider Miyajima deer wild?

	Japanese (n = 276)	International (n = 252)
Wild	61 (22.1%)[a]	34 (13.5%)[a]
Semi-wild	149 (54.0%)	128 (50.8%)
Not wild	40 (14.5%)[b]	76 (30.2%)[b]
I don't know	23 (8.3%)	11 (4.4%)
No answer	3 (1.1%)	3 (1.2%)

[ab]Categories that are significantly different between the two groups (t-test).

Opposition groups

Other stakeholders causing tensions with management officials are opposition groups who oppose the Miyajima deer management policy. Contrary to Miyajima deer management officials, opposition groups do not see the deer as wild. Soon after officials announced the ban on deer feeding, those opposed formed an NPO called *Miyajima no Sika o Itsukushimu Kai*, meaning the Organization that Cherishes Miyajima Deer. On 22 November 2010, the deer management officials and the NPO held a meeting. In response, the NPO sent a letter to Hatsukaichi City officials on 30 November 2010, pointing out discrepancies between the management officials' statement and their understandings of the deer in Miyajima.

The NPO was not convinced by the deer management officials' statement. They pointed out that the officials' statement in the management plan was contradictory and asked them to take responsibility for past actions and reconsider the concept of wild animals. It appears that at the time, the opposition group was not supportive of the approach or narratives that the management officials adopted to achieve their goal. For instance, they were against the sudden termination of feeding, considering such an approach cruel as the Miyajima deer were accustomed to being fed by humans. Importantly, all members of *Miyajima no Sika o Itsukushimu Kai*, which was active from 2010 to 2012 (Hiroshima Prefecture, 2023), were from outside Miyajima, according to a local resident who used to be a member of the group. The organization was originally established to feed the deer on Miyajima, but it was disbanded. Even after the NPO was disbanded, a few members continued to regularly visit Miyajima to bring food to the deer. For management officials, their activity has always been a major obstacle to achieving their management goal. During the deer management meeting the first author attended in 2018, the committee members expressed their concerns as some activists were still regularly visiting Miyajima in a van to feed the deer.

How deer are framed as wild

The root issue surrounding the human–deer conflict is people's different perceptions of the Miyajima deer. Management authorities have emphasized ecosystem conservation and protection of biodiversity. In this context, deer are considered wild. This has been especially emphasized in the Miyajima deer management discourse as there have been constant reminders that the deer in Miyajima are wild through the audio and visual signs. In the Miyajima Deer Protection Management Plan, there is a section explaining how the management authorities view the deer:

> Deer inhabiting Miyajima Island are wild Japanese deer (Cervus nippon) ... Therefore, Miyajima deer should be treated not only as wild under the law but also be protected and managed as biologically wild animals. (Hatsukaichi City, 2008, p. 2)

Discourse analysis revealed two ways in which boundaries were drawn and the deer became wild beings. Spatial boundaries emerged, and most of the discourses framed the deer into nature, usually synonymously described using words such as 'mountain' or 'forest'. Such a spatially oriented discourse was prominent within the management narratives. Reading the texts on signboards revealed the expectation that the deer belong in the mountains:

> The deer on Miyajima Island are (now) used to being fed by humans and they tend to come down to the residential areas for food. As the result, the 'residential deer' are now gradually increasing in those areas. (Original text on one of the signboards found in Miyajima)

Similarly, the official documents of Miyajima Deer Management Plan (Hatsukaichi City, 2014b) indicate that the deer should belong in the mountains, not in the residential area:

> Desirable management is that the deer live wild, which is their natural state, and we can observe them closely in the natural landscape... (Hatsukaichi City, 2014a, p. 1)
>
> In order to return the deer back to the mountain, the management officials set up a feeding station... (Hatsukaichi City, 2014b, pp. 1–2)

The underlying idea denoted in the above statements is that the deer are part of nature, which is signified with mountains and forests. The residential area is a space for humans, and thus the deer do not belong there – it is unnatural for the deer to be frequently present in such a space. The phrase, 'natural state' is indeed problematic for the Miyajima deer as their 'natural state' is likely where they live in the residential area.

Phrases such as 'come/came down' and 'go home' imply that the deer descended (likely) from the mountains, crossed the 'border' and intruded on the 'human space'. Many people believe that mountains are where wild deer have lived. However, this human–deer boundary is merely a product of humans' imagination and a social construction. According to Hayama (2008), deer have actually evolved to live on the plains. In other words, living in the mountains is rather unnatural for them.

In addition to the spatial framing, wild deer are also behaviourally framed as wild. 'Wild animals should not rely on humans for food' is a discourse that has emerged. As Setoguchi (2013) reminded us, the discourse that 'it is inappropriate to feed wildlife' has been established rather recently. In the 1950s, feeding (or provisioning) wild animals was not considered an action to detach them from non-human animals' state of being wild. This is evident in the establishment of monkey parks in Japan. Wild monkey populations were lured to the viewing ground through provisioning and visitors enjoyed seeing the wild monkeys. However, this practice of feeding wild monkeys was criticized and reconsidered in the 1970s as nature conservation became of utmost importance in the world and as reports of crop raiding increased. During this time, the media began reporting on feeding in a negative manner. Humans' perception of wild animals as being free of human influence became widespread globally in the 1970s (Setoguchi, 2013).

Another discourse that frames the deer's behaviour is habituation – that is, getting used to certain situations. Habituated deer are not afraid of human presence. Candea asserted 'The practice of habituation problematizes the boundaries between the wild and the tame' (2010, p. 245). Here, 'tame' is used synonymously with 'domesticated'. 'I don't think Miyajima deer are wild because they follow people' (Tourist, interview 2017) is a quote from an interview conducted on Miyajima Island that illustrates this point. It appears that feeding combined with taming is an important component for people to consider non-human animals wild (or not wild). Exhibiting only one of these behaviours may not adequately make them non-wild creatures.

Transgression of boundaries

The deer on Miyajima Island are positioned as wild animals spatially and behaviourally, but these boundaries can be transgressed. The deer appear in the 'human space' on the island. They are not wary of human presence and even follow people, presumably to obtain food. It is when the boundaries are transgressed that the deer are considered problematic. For instance, the Miyajima deer management plan states:

> Concentration of the deer population in the residential area is causing human–deer conflicts, such as negatively impacting humans' quality of life and harming tourists. In addition, increasing the number of deer is reported to have a detrimental effect on the ecosystem. Hence, we are concerned about the impact that the increased Miyajima deer population would have on Miyajima's nature and landscapes. (Hatsukaichi City, 2014b, p. 1)

This management narrative carries negative connotations of the deer. The issues of hygiene include a bad odour from deer faeces and urine. However, the increase in the number of deer may not be the single cause of the emergence of these problems. One of the interviewees shared a different perspective:

> when I first got to know about Miyajima, the streets were not paved at all. When the deer urinated or defecated, their faeces were absorbed into the ground (soil) and broken down quickly... They did not really smell. Because humans have paved the roads and streets with concrete, it would look dirty if deer's faeces are left there. (Respondent D, 2018)

Moreover, he suggested that management officials may have manipulated the local people's opinion about deer:

> When the management officials distributed the questionnaire, I found it unacceptable because the questionnaire started asking, 'Have you ever had a bad experience with the deer in Miyajima?' Then, this question was followed by a question that asked what kind of troubles did I have ... The questionnaire starts with an assumption that deer are bad ... The questionnaire was used only to back up the management officials' claim that humans' and deer's spaces should be separated. This is why I don't trust this questionnaire data. (Respondent D, 2018)

Managed to be wild

Setoguchi (2013) claimed that the concept of wild (or wildlife, in his words) holds a contradictory dual notion. Although the term wildlife is used regularly in modern society, it is a relatively new word in Japan. The term became generally recognized in the 1930s when animal literature from North America was introduced to Japan, and its meaning has transformed since then. Today, wildlife has come to mean beings that seem to be free of human control, but they are constantly monitored and managed by humans. Kikuchi (2008), for instance, illustrated this duality in the process of reintroducing endangered oriental white storks into the wild. He argued that although the storks are released into the wild, their reproduction is monitored by scientists to avoid inbreeding.

Likewise, the Miyajima case shows the duality of human's imagined wildness and their treatment of the deer. The state of being wild means being away from humans and human influence. Nonetheless, wild deer on Miyajima Island are 'managed' to become wild, which clearly has a human influence. This reveals a double standard regarding the treatment of wild deer.

What becomes unjust in the treatment of Miyajima deer is when accidents are caused by humans. Although the deer are wild in the management officials' eyes, they spend a significant amount of time in the residential areas. Thus, there are likely collisions between the deer and vehicles on the island. The first author has seen deer and bicycles almost collide on multiple occasions. For example, on 4 August 2014, the author saw a male deer whose antler was bent completely. The author had seen the same deer, apparently recently injured, on a previous visit. At that time, his antler was bleeding and other tourists were wondering what to do. The author went to the tourist information centre at the ferry terminal and explained the situation, and was instructed by the lady at the office to leave the deer alone, as the deer are considered wild. Being wild deer on Miyajima Island means that if they are injured, they should be left alone. In some cases, the injured deer are relocated to the enclosed area located opposite the residential area on the island – a place the residents call *ura*, meaning 'the backside'. A retired man, who used to work on the island and now voluntarily gives tours there, said that someone manages the deer, rescues the injured deer and brings them to *ura* because Miyajima is a tourism destination.

This raises a question as to whether intervention on behalf of non-human animals is an ethically correct action. Such a topic has been debated in the context of 'wildness'. While there is no definite answer to this question, pro-interventionists argue that non-human animals deserve equal moral consideration and that humans have a moral obligation to avoid the unnecessary suffering of non-human animals. By contrast, anti-interventionists claim that interventions that could potentially result in negative consequences should be avoided or that non-human animals lack agency and thus do not deserve moral consideration (Torres, 2015).

Borderland politics

A borderland is a space that humans and non-humans share (Wolch *et al.*, 2002). It also describes an ambiguous space in which drawing a clear line in between two contrasting notions is difficult to do (Michel, 1998). It is this borderland that Miyajima's case illustrates. The space where humans interface with deer has been contested as the transgression occurred on humans' imagined boundaries, but not all boundary transgressions lead to contestation. Now, we turn the discussion to the politics of this boundary management.

Yeo and Neo (2010) investigated how people living in the area adjacent to Bukit Timah Nature Reserve (BTNR) in Singapore perceived macaques by analysing discourses, and they found that borderlands are in fact a space for humans. In the study, they were particularly interested in examining how human–macaque conflicts were managed at the borderlands of BTNR. They found that the human–macaque boundary was politicized owing to a fear of negative impacts that could potentially be brought by the monkeys (e.g. crop raiding, disease transmission). Thus, culling the monkeys was legitimized. Moreover, such politicization was extended beyond the 'human space'.

The human–deer borderland also appears to be a place for humans. In Miyajima, the deer became problematized as they crossed this boundary. Note that humans' treatment of the deer was different in the past. As we have shown in the history of Miyajima, the deer were living in the residential area. After the Second World War, when the deer population dropped, Miyajima residents sought to create a landscape with free-ranging wild deer. However, the situation changed over time. Now, the deer roaming around the residential area searching for food from humans are recognized as out of place and measures have been taken to bring them back to 'their space'.

While there is a power at work to restore the human–deer space, once the deer transgress the boundary, the transgression from the opposite direction – from the 'human space' to the 'deer space' – does not seem to destabilize the boundary: 'It is desirable for deer to live in a wild state and humans can closely observe them in nature' (Hatsukaichi City, 2014b, p. 5).

While Miyajima deer are considered wild, and the mountains (i.e. 'natural space') are regarded as their habitat, the 'deer's space' (i.e. nature) is also available for humans (i.e. cultural space). This is an aspect pointed out in the context of dingo–human relationships in Australia (Burns, 2008). With the availability of hiking trails and cable cars in Miyajima, humans can venture into the 'deer's space'. Whether 'wild' space would eventually be invaded by humans remains an open question.

More than just a resource: evidence of deer as active agency

Contrary to management officials' desire for deer to stay in the mountains or forests, the deer do not remain in 'their space' but move from one place to another throughout the island. Humans are the ones who determine the 'imagined' space of deer. Miyajima's deer management policy and decisions have been influenced by the presence and resisting behaviours of the deer to a certain extent. For instance, although Miyajima Town officials established a deer feeding station in the mountains in an attempt to lure the deer there (Hatsukaichi City, 2014b), the deer did not go back to 'their space' as the officials wished them to. Although since the outbreak of the COVID-19 pandemic the number of deer in the residential area has visibly decreased, according to a male resident some deer continue to live close to humans.

Our previous research also indicated that deer are actively searching for food despite the management officials' attempt to prevent them from doing so (Usui and Funck, 2018). For the Miyajima deer, remaining in the residential area has another benefit besides food. A retired male resident reported that deer choose to live near humans because staying close to the human areas protects fawns from predators, such as crows.

The presence of the deer shapes the way people live. *Shikado* (deer gates) is one example and are a unique architectural feature of Miyajima houses (Funatsuki, 2011). It could be a symbolic structure that demonstrates the history of human–deer coexistence in

Fig. 13.4. A mailbox is placed high to prevent deer from eating the mail. (Photo by Rie Usui.)

Miyajima – although the number of houses with *shikado* declined after the deer's population drop (Funatsuki, 2011). Moreover, mailboxes are positioned high above the ground to keep the delivered mail out of the deer's reach (Fig. 13.4).

Deer are also part of people's daily conversations. Sometimes, people refer to a certain individual deer. One local lady talked to a doe: 'Maru-chan's mom, you left your baby again? Where is Maru-chan?' The lady told the first author that this particular deer has a line on her nose. Occasionally, this doe is seen with her fawn. The lady added that the doe does not feed her baby often. On the same day, the author stopped by a café where Miyajima residents or workers sometimes gather. The one small window facing the street was left wide open. The café owner saw a deer walking by and said, 'Ah ... she left her baby again'. He was likely referring to the same doe.

The conversations heard on the island indicate that for some residents the Miyajima deer are not merely deer. They know particular deer and observe how the females raise their offspring, like grandparents watching how a young new mother raises her children. Ishida (2013) stated that the naming of non-human animals in zoo contexts was derived from people's attachment to certain non-human animals. It is apparent that some residents establish attachments to some deer. They know specific deer as individual actors and not simply as one of many deer.

Conclusion

When human and non-human animal conflicts occur, humans decide the fate of the non-human others. On Miyajima Island, the human-centred narratives and views framed the deer as problematic in the space, and they were physically and conceptually contained in the wild space. The image of wild deer is created by drawing two types of boundaries: geographical and behavioural boundaries. Commonly, deer are geographically framed as wild, but humans' image of wild deer is indeed socially constructed. In addition to being confined within a geographical boundary, wild deer are expected to behave in certain ways. Feeding and habituation are two key behaviours that describe wild deer. The common behavioural discourse is that wild deer should not rely on humans to obtain food and thus they should not exhibit tame behaviour.

Once conflicts occur between humans and deer, the deer are negatively represented. Conflicts are typically the result of an imagined boundary transgression. The typical response to such situations is to restore the wild state by managing the deer population. Furthermore, while wildlife is framed spatially as in the wild, human interests are often valued over the deer's interests in 'their space', which clearly demonstrates a power imbalance in the relationship between humans and deer.

What if non-human animals were included among the decision-making members of the committee? The posthumanist political ecology used in this study allowed us to come closer to fair decision-making by considering the deer as important actors and members of the moral community by unravelling the unjust treatment of deer in the process of wilding. Analysing the human–deer space and interactions using the posthumanist political ecological framework enables us to paint a different picture of their coexistence on the island, where deer present in their desired location.

While humans create boundaries to contain deer in the wild space, there is resistance against the deer management practice. The deer are essential actors. They choose to stay in the residential areas and actively seek food from humans, and people build housing structures, such as *shikado*, and mailboxes such that the deer cannot intrude into their personal property. Nonetheless, deer would still eat flyers found in town. Although some residents find such deer behaviours to be a nuisance, the deer form an important topic in residents' daily conversations and in tourists' perceptions of Miyajima. This evidence shows that the presence of deer has, to a certain extent, influenced the existing landscapes of Miyajima Island through contact with the human community. To move forward, we suggest that the management authority recognize Miyajima's unique human–deer relationship as an essential component of cultural heritage.

References

Agency for Cultural Affairs, Government of Japan (2022) A list of Japan's World Heritage. Available at: https://www.bunka.go.jp/seisaku/bunkazai/shokai/sekai_isan/ichiran/ (accessed 29 May 2022).

Barnes, T.J. and Duncan, J.S. (1992) *Writing Worlds: Discourse, Text and Metaphor in the Representation of Landscape*. Routledge, London.

Braidotti, R. (2013) *The Posthuman*. John Wiley & Sons, Cambridge, UK.

Bryant, R.L. (1998) Power, knowledge and political ecology in the third world: a review. *Progress in Physical Geography* 22(1), 79–94. DOI: 10.1177/030913339802200104.

Bryant, R.L. (2000) Politicized moral geographies: debating biodiversity conservation and ancestral domain in the Philippines. *Political Geography* 19, 673–705. DOI: 10.1016/S0962-6298(00)00024-X.

Burns, G.L. (2008) Lines in the sand: an anthropological discourse on wildlife tourism. Doctoral thesis. Murdoch University, Perth, Australia.

Candea, M. (2010) 'I fell in love with Carlos the meerkat': engagement and detachment in human–animal relations. *American Ethnologist* 37(2), 241–258. DOI: 10.1111/j.1548-1425.2010.01253.x.

Castricano, J. (2008) Introduction: animal subjects in a posthuman world. In: Castricano, J. (ed.) *Animal Subjects: An Ethical Reader in a Posthuman World*. Wilfrid Laurier University Press, Ontario, Canada, pp. 1–32.

Catlin, J., Hughes, M., Jones, T., Jones, R. and Campbell, R. (2013) Valuing individual animals through tourism: science or speculation? *Biological Conservation* 157, 93–98. DOI: 10.1016/j.biocon.2012.07.022.

Chugoku Shinbun (1968) *Miyajima no shika: Sukkari ninkimononi [Miyajima deer: Becoming popular]*.

Cohen, E. (2019) Posthumanism and tourism. *Tourism Review* 74(3), 416–427. DOI: 10.1108/TR-06-2018-0089.

Cole, S. (2012) A political ecology of water equity and tourism. A case study from Bali. *Annals of Tourism Research* 39(2), 1221–1241. DOI: 10.1016/j.annals.2012.01.003.

Conover, M.R. (2002) *Resolving Human–Wildlife Conflicts: The Science of Wildlife Damage Management*. CRC Press, Boca Raton, Florida.

Danby, P., Dashper, K. and Finkel, R. (2019) Multispecies leisure: human-animal interactions in leisure landscapes. *Leisure Studies* 38(3), 291–302. DOI: 10.1080/02614367.2019.1628802.

Dashper, K. (2020) Holidays with my horse: Human-horse relationships and multispecies tourism experiences. *Tourism Management Perspectives* 34, 100678. DOI: 10.1016/j.tmp.2020.100678.

Dore, K.M. (2018) Ethnoprimatology without conservation: the political ecology of farmer–green monkey (*Chlorocebus sabaeus*) relations in St. Kitts, West Indies. *International Journal of Primatology* 39(5), 918–944. DOI: 10.1007/s10764-018-0043-9.

Douglas, J.A. (2014) What's political ecology got to do with tourism? *Tourism Geographies* 16(1), 8–13. DOI: 10.1080/14616688.2013.864324.

Escobar, A. (1996) Construction nature: elements for a post-structuralist political ecology. *Futures* 28(4), 325–343. DOI: 10.1016/0016-3287(96)00011-0.

Funatsuki, Y. (2011) *Itsukushima Shin'ezu*. Mediasion, Hiroshima, Japan.

Gaard, G. (2017) Posthumanism, ecofeminism, and inter-species relations. In: MacGregor, S. (ed.) *Routledge Handbook of Gender and Environment*. Routledge, New York, pp. 115–129. DOI: 10.4324/9781315886572.

Grimm, K.E. (2013) Doing 'conservation': effects of different interpretations at an ecuadorian volunteer tourism project'. *Conservation and Society* 11(3), 264. DOI: 10.4103/0972-4923.121029.

Hatsukaichi City (2008) Miyajima region deer protection management planning. Available at: https://www.city.hatsukaichi.hiroshima.jp/uploaded/ attachment/8228.pdf (accessed 31 May 2016).

Hatsukaichi City (2014a) Miyajima deer management plan (2nd term) summary.

Hatsukaichi City (2014b) Miyajima region deer protection management planning (second term). Available at: https://www.city.hatsukaichi.hiroshima.jp/uploaded/attachment/8229.pdf (accessed 27 January 2023).

Hatsukaichi City (2023) Miyajima raitōshasū ichiranhyō [Miyajima incoming tourists chart]. Available at: https://www.city.hatsukaichi.hiroshima.jp/uploaded/attachment/69236.pdf (accessed 8 November 2023).

Hayama, S. (2008) *Wildlife Issues*, 4th edn. Chijinshokan, Tokyo.

Hiroshima Prefecture (2023) NPO hōjin jōhō saito [Information about NPO]. Available at: https://www.pref.hiroshima.lg.jp/site/npo/1271204301612.html (accessed 27 January 2023).

Hobson, K. (2007) Political animals? On animals as subjects in an enlarged political geography. *Political Geography* 26(3), 250–267. DOI: 10.1016/j.polgeo.2006.10.010.

Iketani, K. (2008) Haijo no ronri kara kyōzon no ronri e [From exclusion to coexistence]. In: Hayashi, Y., Mori, Y., Akishinonomiya, A., Iketani, K. and Okuno, T. (eds) *Yasei to Kankyo [Wildlife and Environment]*. Iwanamishoten, Tokyo, pp. 296–319.

Ishida, O. (2013) 'Fureai' to osewa ['Interaction' and care]. In: Ishida, O., Hamono, S., Hanazono, M. and Setoguchi, A. (eds) *Japanese Attitudes toward Animals: A History of Human–Animal Relations in Japan*. University of Tokyo Press, Tokyo, pp. 209–225.

Kikuchi, N. (2008) Ambiguous 'wild' in the case of the reintroduction project of the oriental white stork. *Journal of Environmental Sociology* 14, 86–100.

Kline, C. (2022) Socialising animal-based tourism. In: Higgins-Desbiolles, F., Doering, A. and Bigby, C.B. (eds) *Socialising Tourism: Rethinking Tourism for Social and Ecological Justice*. Routledge, New York, pp. 195–213. DOI: 10.4324/9781003164616.

Lloro-Bidart, T. (2017) A feminist posthumanist political ecology of education for theorizing human-animal relations/relationships. *Environmental Education Research* 23(1), 111–130. DOI: 10.1080/13504622.2015.1135419.

Michel, S.M. (1998) Golden eagles and the environmental politics of care. In: Wolch, J. and Emel, J. (eds) *Animal Geographies: Place, Politics, and Identity in the Nature-Culture Borderlands*. Verso Books, London, pp. 162–187.

Miyajima Tourism Association (2017) Home. Available at: http://www.miyajima.or.jp/amsar/index.html (accessed 30 May 2022).

Mostafanezhad, M., Norum, R., Shelton, E.J. and Thomston-Carr, A. (2016) Introduction. In: Mostafanezhad, M., Norum, R., Shelton, E.J. and Thompson-Carr, A. (eds) *Political Ecology of Tourism: Community, Power and the Environment*. Routledge, New York, pp. 1–21. DOI: 10.4324/9781315717227.

Nepal, S., Saarinen, J. and McLean-Purdon, E. (2018) Introduction: political ecology and tourism-concepts and constructs. In: Nepal, S. and Saarinen, J. (eds) *Political Ecology and Tourism*. Routledge, New York, pp. 1–15.

Neumann, R.P. (2004) Moral and discursive geographies in the war for biodiversity in Africa. *Political Geography* 23(7), 813–837. DOI: 10.1016/j.polgeo.2004.05.011.

Nyhus, P.J. (2016) Human–wildlife conflict and coexistence. *Annual Review of Environment and Resources* 41(1), 143–171. DOI: 10.1146/annurev-environ-110615-085634.

Ono, K. (1962) *Shinto: The Kami Way*. Tuttle Publishing, Tokyo.

Philo, C. and Wilbert, C. (2000) *Animal Spaces, Beasty Places: New Geographies of Human–Animal Relations*. Routledge, New York.

Ponting, H.G. (ed.) (2005) Enourawan to miyajima [Enoura Bay and Miyajima Island]. In: *Lotus-Land Japan*. Academic Library, Japan, pp. 302–323.

Ramsar (2012) Miyajima. Available at: https://rsis.ramsar.org/ris/2056 (accessed 27 January 2023).

Richards, G. (2018) Cultural tourism: a review of recent research and trends. *Journal of Hospitality and Tourism Management* 36, 12–21. DOI: 10.1016/j.jhtm.2018.03.005.

Setoguchi, A. (2013) Perception about 'wild'. In: Ishida, O., Hamano, S., Hanazono, M. and Setoguchi, A. (eds) *Japanese Attitudes toward Animals: A History of Human–Animal Relations in Japan*. University Tokyo Press, Tokyo, pp. 145–156.

Soper, K. (2012) The humanism in posthumanism. *Comparative Critical Studies* 9(3), 365–378. DOI: 10.3366/ccs.2012.0069.

Stonich, S.C. (1998) Political ecology of tourism. *Annals of Tourism Research* 25(1), 25–54. DOI: 10.1016/S0160-7383(97)00037-6.

Sundberg, J. (2011) Diabolic *Caminos* in the desert and cat fights on the Río: a posthumanist political ecology of boundary enforcement in the United States–Mexico borderlands. *Annals of the Association of American Geographers* 101(2), 318–336. DOI: 10.1080/00045608.2010.538323.

Thomsen, B., Thomsen, J., Copeland, K., Coose, S., Arnold, E. *et al*. (2021) Multispecies livelihoods: a posthumanist approach to wildlife ecotourism that promotes animal ethics. *Journal of Sustainable Tourism* 31(5), 1195–1213. DOI: 10.1080/09669582.2021.1942893.

Tomassini, L., Bertella, G., Grasso, C. and Lenzi, C. (2022) The space of animal justice in wildlife sanctuaries: a posthuman perspective. *Journal of Ecotourism* 1–18. DOI: 10.1080/14724049.2022.2122480.

Torres, M. (2015) The case for intervention in nature on behalf of animals: a critical review of the main arguments against intervention. *Relations* 3(1), 33–49. DOI: 10.7358/rela-2015-001-torr.

Treves, A. and Karanth, K.U. (2003) Human-carnivore conflict and perspectives on carnivore management worldwide. *Conservation Biology* 17(6), 1491–1499. DOI: 10.1111/j.1523-1739.2003.00059.x.

Tsing, A.L. (2005) *Friction: An Ethnography of Global Connection*. Princeton University Press, Princeton, New Jersey. DOI: 10.1515/9781400830596.

Turner, M.D. (2015) Integrating politics and ecology through mixed methods. In: Bryant, R.L. (ed.) *The International Handbook of Political Ecology*. Edward Elgar Publishing, Cheltenham, UK, pp. 530–544. DOI: 10.4337/9780857936172.

UNESCO (2022) UNESCO World Heritage List: Itsukushima Shinto Shrine. Available at: http://whc.unesco.org/en/list/776 (accessed 30 May 2022).

Usui, R. and Funck, C. (2017) Not quite wild, but not domesticated either: contradicting management decisions on free-roaming sika deer (*Cervus nippon*) at two tourism sites in Japan. In: Limar, I.B. and Green, R.J. (eds) *Wildlife Tourism, Environmental Learning and Ethical Encounters, Geoheritage, Geoparks and Geotourism*. Springer International Publishing, Cham, Switzerland, pp. 247–261. DOI: 10.1007/978-3-319-55574-4.

Usui, R. and Funck, C. (2018) Analysing food-derived interactions between tourists and sika deer (*Cervus nippon*) at Miyajima Island in Hiroshima, Japan: implications for the physical health of deer in an anthropogenic environment. *Journal of Ecotourism* 17(1), 67–78. DOI: 10.1080/14724049.2017.1421641.

UNESCO World Heritage Convention (2022) Available at: https://whc.unesco.org/en/ (accessed 29 May 2022).

Witter, R. (2013) Elephant-induced displacement and the power of choice: moral narratives about resettlement in Mozambique's Limpopo National Park. *Conservation and Society* 11(4), 406. DOI: 10.4103/0972-4923.125756.

Wolch, J., Emel, J. and Wilbert, C. (2002) Animal geographies. *Society & Animals* 10(4), 407–412. DOI: 10.1163/156853002320936881.

Woodroffe, R., Thirgood, S. and Rabinowitz, A. (2005) *People and Wildlife: Conflict or Coexistence*? Cambridge University Press, Cambridge, UK. DOI: 10.1017/CBO9780511614774.

Yeo, J.H. and Neo, H. (2010) Monkey business: human–animal conflicts in urban Singapore. *Social & Cultural Geography* 11(7), 681–699. DOI: 10.1080/14649365.2010.508565.

14 Conclusion. Shifting Animal Abuse in Heritage Tourism to the Paradigm of Building Multispecies Communities

Leonardo Garavito-González*, Jorge Iván Ruiz Barrera , Estephania Sepúlveda Perdomo, Tomás Arias Roncancio and Mateo Nicolás Rico Medina
Universidad Externado de Colombia, Bogota, Colombia

Abstract
This chapter presents a final balance of the panorama offered by the range of investigations compiled in this book. The mistreatment and commodification of animals at tourism and heritage sites on a global scale appears to be a constant, often framed within the context of anthropocentrism. Nonetheless, the authors invite us to question the appointment of animals as heritage and also the traditions and cultural or tourist practices that involve the harmful treatment of animals.

Shifting Animal Abuse in Heritage Tourism to the Paradigm of Building Multispecies Communities

Framed in anthropocentrism, the mistreatment and commodification of animals in tourism and heritage sites on a global scale seem to be a constant. This doesn't sound like much of a novelty in a world in which the paradigm of human exceptionalism naturalizes relationships of control and ownership between *Homo sapiens* and non-human animals. The management of tourism and heritage is framed in some form of capitalist logic that investigations like the ones presented in this book help make visible, document and reflect on.

To question and try to transform the established order is not an easy task. The beliefs and practices around a supposed moral superiority of mankind are present in various cultures and social groups on the planet and have been normalized by centuries or even millennia. However, for those who recognize the ethical dilemmas inherent to such notions, this type of research offers light and some clues as to what we should change and how we could do it. We need to move towards environments in which the intrinsic value of other animals is recognized beyond their instrumental abilities and where their needs are respected and their interests better understood. The works that make up this book lead us into a challenging but possible transition towards the construction of communities and multispecies cultures that are less unfair and less speciesist.

This chapter presents a final balance of the panorama offered by the range of investigations compiled in this book. To conduct an analysis of their contents, a meaning matrix was built considering aspects such as: (i) location of the case studies, types of tourism and animals involved; (ii) objectives and central research questions;

*Corresponding author: leonardo.garavito@uexternado.edu.co

 Tourism, Heritage and Commodification of Non-human Animals (eds A. López López *et al.*)
DOI: 10.1079/9781800623309.0014

(iii) theoretical framework and methodological strategies; and (iv) main conclusions and further study proposals. A number of common trends and outstanding particularities became evident, illustrating the diversity and relevance of social research that links tourism, heritage and the commodification of animals from a critical perspective. Researchers interested in starting or deepening this line of inquiry will find in these texts theoretical and conceptual elements and useful methodological examples.

On the study cases: charismatic animals exploited by cultural tourism

The various places and cases analysed have at least three characteristics in common. First, they are tourist destinations with heritage attractions that, in most cases, are cultural. Second, animals play a prominent role in driving tourist demand for the destination, and in this sense charismatic animals are the main protagonists. Third, the predominant relationships between humans and animals in these places are characterized by the domination, exploitation and commodification of the latter by the former.

The central type of tourism in most of the cases in this book corresponds to cultural tourism, particularly intangible heritage tourism that includes practices such as bullfighting in Mexico (Chapters 7 and 8) and a ritual/sports version of bullfighting in India (Chapter 6) or gastronomic traditions in Mexico (Chapter 5), or the use of animals as tourist symbols, such as the cases of axolotls in Mexico (Chapter 10), panda bears in China (Chapter 12) and the case of the eagles in an Indigenous reservation in the USA (Chapter 9). The analysis of zoos in Europe and America (Chapter 3), and of cats (Chapter 11) and deer (Chapter 13) in Japan, represent a variety of cultural tourism in which animals are an integral part in the production of heritage landscapes, while the case of palm civets in Indonesia (Chapter 4) is an example of agrotourism.

The predominant type of animals studied in this book is charismatic, that is, animals that have popular appeal and attract the attention of tourists, generating feelings of empathy or admiration. Some examples addressed in this book are eagles, axolotls, cats, panda bears and deer, whose contemplation by tourists as well as the emotional 'pull' of the animal play a central role. In contrast, other animals have a more active role, such as bulls in bullfighting, or a more passive role as part of the production process of a commodity, such as the captive wild palm civets do relative to gourmet coffee. In the tourist and human entertainment business, such human–animal relationships mediated by use and abuse, reification and commodification have become standard. A relevant exception seems to be the relationship between eagles and Native Americans where, as the analysis suggests, the active rejection of their commodification and human predominance can serve as an example of good practices in the field. We shall go back to this later.

On the objectives and questions: the analysis of the practices and meanings of animal abuse

The works that make up this book tend to address two types of subject matter: on the one hand, the use of animals and the practices of different types of actors (tourist service providers, tourists, authorities) in tourist or heritage destinations, and on the other, the main meanings, arguments or discourses that inform them.

As for practices, Chapter 4 explicitly illustrates the ways in which civets and Indigenous people in Bali are exploited for the entertainment of tourists and economic gain. Chapter 5, which focuses on Sonora Market in Mexico City, critically examines the practices of commodification of animals, as well as the ethical problems derived from them. Chapter 9 examines tourist and local practices associated with ancestral knowledge and eagles during visits to the sanctuaries of Indigenous communities. Chapter 10 explores the objectification of the axolotl into a brand and product for tourist consumption in Mexico. And Chapter 11 investigates the use of cats in Japanese heritage sites to attract more tourists. On balance, three perspectives can be identified: first, the denunciation of cruel or abusive practices towards animals, as in the case of Bali or Sonora Market; second, the consideration of strategic tourism practices that, although they exploit animals, do not necessarily commit acts of cruelty or mistreatment, as with the axolotl in Mexico or cats in Japan; and third, the identification of good practices characterized by respect and appreciation of animals, as in the study on Native Americans and eagles.

As for meanings, Chapter 3 reviews the predominant arguments in the discussion about zoos as cultural institutions and their ethical standing. Chapters 7 and 8 review the articulation of the tourist and patrimonial discourses about bullfighting in Mexico, as well as its dissonances with the discourses on animal ethics, the arguments in favour of its prohibition and the negative perceptions on the part of some sectors of the local population and of tourists. Chapter 13 explores the conceptual conflict inherent in the construction of human space as opposed to wilderness in the relations between humans and deer in Japan. To a large extent, these analyses attest to the emergence of critical perspectives on the traditional ideas about human superiority or exceptionality, and the promotion of a less hierarchical and more horizontal human–animal relationship.

Finally, two chapters analyse practices as well as discourses. Chapter 6, on the ritual use of bulls in India, contrasts the cultural practices involved in such ceremonies with the arguments of animal law that question these traditions. Chapter 12 analyses the social construction of the ‘cute/endearing’ heritage in the practices at parks and breeding centres for panda bears and the various documents and advertising images derived from them.

On the theoretical-methodological framework: the posthumanist gaze and qualitative research

The central theoretical referent of this book is, undoubtedly, the posthumanist proposal, expressed in the form of three fundamental ideas for the interpretation of several of the case studies. First, a critique of anthropocentrism, which places humans at the centre of the universe and regards other forms of life and elements of nature as their property and mere satellites at their disposal. Second, a critique of speciesism that denounces the various types of discrimination living beings suffer at the hands of humans on account of their belonging to a different species. Such types of discrimination vary in intensity according to the levels of empathy and identification accorded by humans. Domestic or charismatic species enjoy the highest level, farm animals an intermediate one, and wild animals and insects the lowest level. Third, a critique of the human exceptionalism paradigm, according to which humans are supreme among living beings on account of their cognitive abilities (from a secular standpoint) or because they are the masters of Creation (from a religious perspective). For either of these reasons, the care and management of nature are entrusted to humans.

Among the volumes that address posthumanism, the work of Cohen (2019) stands out for the links it establishes with tourism. This author is cited in five of the chapters of this book. Other important authors and proposals, featured in at least three of the chapters, are Fennell (2012a, b) on the links between animal ethics and tourism and Regan (2004) on animal rights issues. Finally, other references concerned with posthumanism to an extent and used in at least one of the chapters are O'Neill (1997), Sundberg (2011), Haraway (2013), Evans (2015), Dale *et al.* (2016), Donaldson and Kymlicka (2016), Guia (2020), Horta (2020), Kline (2021) and López and Quintero (2021).

Similarly, specific theoretical references support specific research problems. In dealing with the concept of heritage, Chapters 7 and 8 on bullfighting in Mexico cite the works of Ashworth and Tunbridge (1996) and McDowell (2008). Chapter 9, on eagles and the cultural values of North American indigenous communities, features Zeppel (2010) and Walker and Moscardo (2016). Finally, Chapters 11 and 12 on cats in Japan and panda bears in China include references to Ahmed (2004), Lukacs (2015) and Miyamoto (2016) pertaining to the affective economies of the cute and the endearing.

In terms of methodology, the qualitative approach is common to all the chapters. This methodology is congenial to this field of research in which the problems must be understood in their particular contexts and therefore the meanings and practices displayed by the various actors involved must be emphasized. The collection techniques, from most to least used, include documentary reviews, interviews, direct observation, ethnography, analysis of digital content on social networks and surveys.

On the conclusions and research horizons: between abuse and a multispecies community

The conclusions derived from these chapters can be classified into three categories, which, in turn, open up opportunities for further research in the field. First, ethical conflicts derived from highly unbalanced power relations between humans and other animals; second, cultural tensions between various traditional or heritage practices and animal rights; and third, a series of questions regarding the agency and responsibility of tourists, tourism service providers, communities, authorities and even non-human animals themselves.

In the light of animal and posthumanist ethics, animals associated with various tourist destinations and heritage attractions tend to be immersed in highly unequal power relations *vis-à-vis* their human counterparts (tourists, service providers and authorities, among others). In this sense, many of the tourist and heritage narratives based on the value of their cultural and traditional elements, material or immaterial, are transformed into stories of animal suffering, exploitation and control. Undoubtedly, human interests in economic profit (coffee in Indonesia, Sonora Market in Mexico), the conservation of cultural practices (bullfighting in Mexico and India), the control of space (deer in Japan) and tourism development (cats in Japan, zoos in general), among other things, are superimposed on the needs of other animals, with no regard for ethical and legal considerations.

In some instances, the abuse or mistreatment of animals is not so obvious or direct, but nevertheless needs to be examined from an ethical perspective. The use of animal images as trademarks (as in the case of axolotls in Mexico or pandas in China) is a case in point. It could be argued that at least part of the income generated by the use of their image can be invested in promoting their conservation, but to what extent does this justify this type of animal exploitation for economic gain?

The analysis of pandas in China and cats in Japan illustrates a growing tendency to use animals and their image in tourist and heritage destinations to generate added value. This is reflected in novel concepts such as 'kawaiification' or 'the heritage of beauty' that point to an expansion of the tourist and heritage fields towards the exploration of the economies of affection and emotions. Thus, animals are reified to satisfy people's desires and consumption needs at the expense of their own desires and needs, which may imply restrictions on their mobility and changes in their diet and their socialization patterns, to name a few impacts.

On another level are the conflicts between cultural rights and animal rights or, in other words, the imbalance between beliefs and cultural or tourist practices in relation to animal welfare. In this regard, Chapter 7, which explores bullfighting in Mexico, states that UNESCO (United Nations Educational, Scientific and Cultural Organization), the highest international authority on cultural and heritage issues, allows various violent practices against animals whenever they are considered heritage, thereby legitimizing and prolonging their existence.

In this sense, Chapter 8, which focuses on bullfighting, directly proposes that not everything that has been recognized as heritage is worth preserving, especially when the mistreatment of animals for human enjoyment is involved. There is a growing social demand in favour of the recognition, strengthening and more rigorous application of animal rights. On the other hand, Chapter 4 on civet coffee agrotourism offers a relevant caveat about the dangers of corrupting traditions and cultural heritage.

These conflicts open up a space for reflection about the agency and responsibility of the various actors involved in these processes. As discussed in Chapter 5 on Sonora Market, there is an important opportunity for tourists to reflect on their role and to value the existence and self-interests of the animals present in the destinations they visit. Indeed, the works that analyse the point of view of tourists or their comments on social networks – bullfighting in

Mexico, civets in Indonesia – reveal a growing opinion and sensitivity in favour of animal welfare and against animal exploitation.

To a large extent, tourism service providers can play a leading role in this transformation. The case of the eagles and the Native American community (Chapter 9) is a great example in this regard. In this case, the community assumes the challenge of offering an educational and awareness-raising experience to its visitors, which points towards the recognition of the value of these animals not as a simple source of entertainment or economic gain, but as an integral part of the network of life and as symbols of a deep bond between humans and nature. Along this same path, Chapter 13 on deer in Japan concludes that recognizing these animals as important actors and active members of a multispecies moral community could generate fairer decision-making processes for the management of tourist and heritage destinations.

In conclusion, some questions and further research proposals and horizons are worth highlighting. First, the relevance of extensive studies on the welfare of animals. In line with the criticism of zoos, the authors invite us to examine sanctuaries as a new institution for animal care. Second, the need for expanded research on multispecies relationships in tourist and heritage sites from a posthumanist perspective. In particular, authors invite us to question the appointment of animals as heritage and to further the research on traditions and cultural or tourist practices that involve the harmful treatment of animals. Third, and extending the previous point, they propose a revision of the concept of heritage through questions such as 'whose heritage, and for whose benefit?' Fourth and finally, they encourage building programmes and receptive contexts for multispecies encounters outside of a mercantile, instrumental and inequitable framework, such as the one that currently dominates the management of tourist and heritage sites.

References

Ahmed, S. (2004) Affective economies. *Social Text* 22(2), 117–139. DOI: 10.1215/01642472-22-2_79-117.

Ashworth, G. and Tunbridge, J. (1996) *Dissonant heritage: The Management of the Past as a Resource in Conflict*. Wiley, Chichester, UK.

Cohen, E. (2019) Posthumanism and tourism. *Tourism Review* 74(3), 416–427. DOI: 10.1108/TR-06-2018-0089.

Dale, J.P., Goggin, J., Leyda, J., McIntyre, A.P. and Negra, D. (2016) The aesthetics and affects of cuteness. In: Dale, J.P., Goggin, J., Leyda, J., McIntyre, A.P. and Negra, D. (eds) *The Aesthetics and Affects of Cuteness*. Routledge, London, pp. 11–44. DOI: 10.4324/9781315658520.

Donaldson, S. and Kymlicka, W. (2016) Comment: between wild and domesticated: rethinking categories and boundaries in response to animal agency. In: Bovenkerk, B. and Keulartz, J. (eds) *Animal Ethics in the Age of Humans: Blurring Boundaries in Human-Animal Relationships*. Springer, Cham, Switzerland, pp. 225–242. DOI: 10.1007/978-3-319-44206-8.

Evans, W. (2015) Posthuman rights: dimensions of transhuman worlds. *Teknokultura* 12(2), 373–384. DOI: 10.5209/rev_TK.2015.v12.n2.49072.

Fennell, D.A. (2012a) Tourism, animals and utilitarianism. *Tourism Recreation Research* 37(3), 239–249. DOI: 10.1080/02508281.2012.11081712.

Fennell, D.A. (2012b) *Tourism and Animal Ethics: Contemporary Geographies of Leisure, Tourism and Mobility*. Routledge, London and New York.

Guia, J. (2020) Conceptualizing justice tourism and the promise of posthumanism. *Journal of Sustainable Tourism* 29(2–3), 503–520. DOI: 10.1080/09669582.2020.1771347.

Haraway, D.J. (2013) *When Species Meet*, 3rd edn. University of Minnesota Press, New York.

Horta, O. (2020) Qué es el especismo? *Devenires* 21(41), 163–198.

Kline, C. (2021) Socialising animal-based tourism. In: Higgins-Desbiolles, F., Doering, A. and Chew Bigby, B. (eds) *Socialising Tourism: Rethinking Tourism for Social and Ecological Justice*. Routledge, London, pp. 195–213. DOI: 10.4324/9781003164616.

López, A. and Quintero, G. (2021) Animal dark tourism in Mexico: bulls performing their own slaughter. In: Rickly, J. and Kline, C. (eds) *Exploring Non-Human Work in Tourism*. De Gruyter Oldenbourg, Berlin, pp. 69–82. DOI: 10.1515/9783110664058.

Lukacs, G. (2015) The labor of cute: net idols, cute culture, and the digital economy in contemporary Japan. *Positions. Asia Critique* 23(3), 487–513. DOI: 10.1215/10679847-3125863.

McDowell, S. (2008) Heritage, memory and identity. In: Howard, P. and Graham, B. (eds) *The Ashgate Research Companion to Heritage and Identity*. Routledge, London and New York, pp. 37–53. DOI: 10.4324/9781315613031.

Miyamoto, K. (2016) ネコノミクスの経済効果 [Economic effects of nekonomics]. Available at: https://www.kansai-u.ac.jp/global/guide/pressrelease/2015/No44.pdf (accessed 9 November 2023).

O'Neill, O. (1997) Environmental values, anthropocentrism and speciesism. *Environmental Values* 6(2), 127–142. DOI: 10.3197/096327197776679121.

Regan, T. (2004) *The Case for Animal Rights*, 2nd edn. University of California Press, Berkeley and Los Angeles, California.

Sundberg, J. (2011) Diabolic caminos in the desert and cat fights on the Río: a posthumanist political ecology of boundary enforcement in the United States–Mexico borderlands. *Annals of the Association of American Geographers* 101(2), 318–336. DOI: 10.1080/00045608.2010.538323.

Walker, K. and Moscardo, G. (2016) Moving beyond sense of place to care of place: the role of indigenous values and interpretation in promoting transformative change in tourists' place images and personal values. *Journal of Sustainable Tourism: Sustainable Tourism and Indigenous Peoples,* 24(8–9), 1243–1261. DOI: 10.1080/09669582.2016.1177064.

Zeppel, H. (2010) Managing cultural values in sustainable tourism: conflicts in protected areas. *Tourism and Hospitality Research* 10(2), 93–115. DOI: 10.1057/thr.2009.28.

Index

Note: Page numbers in **bold** type refer to **figures;** Page numbers in *italic* type refer to *tables*

Abenomics 140
aestheticism/aestheticization 158, 160
affective/affection 158
  atmospheres 159
  economics 160–161
  labour 159
aggressive behaviours 42
agrotourism 49, 52
Aguascalientes 86, 93
  bullfighting in **107**
Ahmed's theory of affective economics 158
Ambystoma mexicanum 133
animal
  behaviour 38
  in captivity 42
  carcasses 62
  cause 78
  cruelty 75
  dark tourism 27
  -derived gastronomic heritage 48–49
  ethics 74, 88
    discourses 95
    and tourism 187
  sacrifices 75
  sanctuaries 43–44
  tourism 2, 26, 69–70
  -tourist interactions 13
  welfare and rights school 78
animal abuse 105
  multispecies communities 185–189
  practices and meanings of 186–187
  see also bullfighting
Animal Heritage and Tourism 12
animal-human encounter 2
animalist ideology 94–95
animality 2, 8
Animal Protection Law of Sonora 95
animal rights 35, 187–188
  activists 80
  and animal welfare, concepts 76
  and cultural rights 80
  groups 70
  recognition 43
  and species justice 113
Animal Welfare Law of Guerrero 94–95
Anthropocene 3, 26
anthropocentrism 12, 102–103, 185, 187
anti-ban campaign 82
antihumanism 8
anti-interventionists claim 179
anti-natalism 9
anti-place 68–69
anti-speciesist
  feminism 10
  posthumanism 12
apolitical moralism 74–75
argumentation 41
artificial intelligence 8
artistic discourse 93–94
Asian palm civet *(Paradoxurus hermaphroditus)* 47, **50**
Association of Zoos and Aquariums (AZA) 39
atrocity heritage 62–63
Australian Coral Reef 26
authentic agricultural heritage 52
'authentic' cultural tradition 92
authenticity concept 49
Autonomous Communities 91
Autonomous Community of Catalonia 80–81
autonomy and resistance 143

Aviary 121
AWBI (Animal Welfare Board of India, a statutory body set up by an act of Parliament) 75
axolotl *(Ambystoma mexicanum)* **132**
- -bus **135**
- Chignahuapan
  - themed *Zoopolis* 132–134
  - theming 130–131
- conservation 131–132
- contested space and the teleological metamorphosis 134–136
- Environmental Management Unit (UMA) 136
- exhibition and commercialization 128–129
- local community role 129
- and local politics **133**
- MUMAX 136
- *Pueblos Mágicos* programme 129–131
- specimens 132

'Bali Cat Poo Chino' 54, **55**
Balinese Kopi Luwak Agrotourism
- agrotourism 49
- animal-derived gastronomic heritage 48–49
- authentic agricultural heritage 52
- civet
  - adoption, as heritage icon 53–55
  - exploitation as unethical element of heritage 55–57
- coffee production and consumption 51–52
- methods 48
- natural capital and natural heritage 48
- natureculture framework 49–50
- as tastescape of Bali 52–53
- tour operators and tourists, trust between 58
- traditional Balinese landscape, desire to consume 50–51
- traditional narratives 57

bibliographic searches on animal geography 61–62
biobanks 39
biocentrism 8
biodiversity 158
biology-related learning 37
'biophilia hypothesis' 49
biotic and abiotic communities 8
Bitchū Matsuyama Castle 140–141, 147–149, **150–152**
borderland politics 179–180
boundaries, transgression of 178
breeding programmes 42
Buenos Aires Zoo 43
Bukit Timah Nature Reserve (BTNR) 179
bullfighting 2, 75, 77
- in Aguascalientes **107**
- analysis as dissonant heritage 88–90
- bullfighter's identity 94
- as cultural-touristic element 102–103
- culture 86
- between heritage and animal commodification 100–101
- heritage ideology 91–94
- humanist and posthumanist perspectives 101–103
- methodology 103–104
- official proclamation of 99–100
- at San Marcos National Fair (SMNF) 99–109
- as tourist attraction 104
- tourists' interests in SMNF activities **106**

bullrings 92
bull-taming 78

Cambridge Declaration on Consciousness, The 9–10
capitalist/capitalization 157–158
- exploitation 27
- logic 2
- -speciesist 5

Capitalocene 3
captive/captivity 40–42, 55
- and animal rights 41–43
- breeding 41

Cartesian argument 8
'castle in the sky' 147
- *see also* Bitchū Matsuyama Castle

cat boom 141
charisma 158
Chengdu International Finance Square (Chengdu IFS) 161, **162**
Chengdu Research Base of Giant Panda Breeding 161
Chicago Zoological Park 43
Chignahuapan 128
- anthropocentrism inter-specific relations 129
- arrival of tourism in 136
- axolotl as protector of 128, **134**
- 'brand monumental letters' 134
- Morena (National Regeneration Movement) Party 133
- themed *Zoopolis* 132–134
- theming 130–131
- for tourism consumption 129
- urban space of 133–134

civet
- adoption, as heritage icon 53–55
- exploitation as unethical element of heritage 55–57
- housing 55

cockfighting 77
coffee
- enthusiasts 52
- lovers 52
- production and consumption 51–52

colonial economic ventures 77
colonialist imperialism 35
commodification 141–142, 161, 186
- of bulls 101

feline saviours 141–142
common heritage of humanity 5
conservation 39–41
programmes integration 43
contextual affordability 53
contraception 42
Convention on Biological Diversity 41
Copenhagen Zoo 43
cosmism 8
'cruel bull-abusing contact game' 80
cultural
authentication 2
consumption 85
diversity 5
and ecological stress 51
education 3
heritage 48, 53, 57, 62, 81, 85, 88–89
of Aguascalientes 99
identity 53
or moral distance 89
practices 63, 88–89
reawakening 121
relativism 12, 77
rights 81, 188
tourism 62, 86, 168, 186
uplifting 125
cuteification strategies 150–151
cuteified heritage, feline saviours 142
cuteness-based heritage functions 4
cynegetic tourism 12, 26

Danube Delta in Romania 26
dark tourism 63, 88
space 69
see also Sonora market, animal trade and dark tourism in
deer
as active agency 180–181
as wild 177–178
De la Cruz-Badiano Codex 68
democratic formal justice, *jallikattu* 76
depeasantization 129
dingo-human relationships 179–180
discourse analysis, Miyajima island 177
dissonance 88
dogfighting 77
domestication 38–39, 143

ecocentrism 8
ecoethics and bioethics 8
ecofeminism 12, 171
ecological-scenic value 14
economistic discourse 93
eco-systemic equilibrium 9
educational character, zoos 36–38
electronic-word-of-mouth advertisement 51
Elephant Sanctuary in Tennessee 43
emotional attachment 158
endangered species of AZA 39
enrichment 153
environmental enrichment 42
environmental ethics 12
environmental imbalance 9
Environmental Management Unit (UMA) 131–132, 136
environmental practices 170
esotericism 64–65
ethics/ethical
animal 74, 88
discourses 95
and tourism 187
and animal welfare 113
awareness 144
conflicts 188
consumerism 58
environmental 12
inclusion 81
justification 36
management decision-making 170
tourism geography and animal ethics 61–62
European Association of Zoos and Aquariums 43
European Convention on Human Rights (ECHR) 81
European-style zoo 35
exhibition and commercialization 128–129

faecal commodification 53, 57
family outing 36
fantasized assimilation 53
farmers' economic distress 79
Fazhan Village 163–165, **164**
Federal Entity 92–93
feline saviours
Bitchū Matsuyama Castle 147–149, **150–152**
cats, liminality and stray agency 142–144
commodification 141–142
cuteified heritage 142
free-roaming populations 143–144
heritage tourism 142
in Japanese Heritage Tourism 144
kawaii 141–142
Kishi station 144–147
nekonomics 141
Sanjūrō 147–149
success 141
feminist epistemologies 4
FIAPO (Federation of Indian Animal Protection Organisation) 75
fiesta brava 99
flat ontology 123
Forbidden City 142
ForceAtlas2 64
foreign influence 79

Foucauldian analysis 62
Foundation for Advice and Action in the Defence of Animals (FAADA) 12
Fray Bentos Industrial Landscape 27
freedom 41–42
free-roaming populations, feline saviours 143–144

Gephi software 64
Giant Panda National Park in Yingjing County 162–163
Global Code of Ethics for Tourism 69
global commodification 74–75
Global Federation of Animal Sanctuaries 43
globalization of heritage 100
gourmet culinary heritage 27

habituated deer, Miyajima island 178
Haraway's concept of lively capital 158
Head-Smashed-In Buffalo Jump 27
Hello Kitty cartoons 141
Hello Kitty-themed World Heritage exhibition 142
heritage
- and animal commodification, bullfighting 100–101
- -based knowledge and cultural perspectives 123
- concept 187
- construction 88
- critical approaches to 10–12
- and cultural tourism, *jallikattu* in 74
- and cultural value 75
- discourse 93
- tourism 27, 85–86, 143
- tourism, feline saviours 142
- tourist attractions, zoos as 34

heritagization 11, 88, 94
Hispanic American Bullfighting Cultural Promotion (HABCP) 104
human-animal relationships 88, 94, 125, 159
human-animal rhythms 159
human-deer interaction
- borderland 179
- conflicts 170–171
- history 172–173
- management *173*, 173–174
- *see also* posthumanist political ecology approach

human/humanism 8–10, 101–102
- entertainment business 186
- exceptionalism 8, 49, 89, 113, 159, 187
- humanist and posthumanist perspectives 101–103
- and non-human animals 170
- oppression 143
- rights 35
- space 179
- -wildlife conflict 169

hybridized modes of tourism 123

Idakiso Station 146
inbreeding in zoos 41
inconsistency dissonance 89
India and *jallikattu* 77–78
Indigenous-led resurgence 125
Indigenous Potawatomi cultural interpretation 123
Indigenous resurgence and grounded normativity **125**
inevitability of conflict 89
inherent value and welfare 123
institutional heritage documents 88
instrumentalist mentality 129
Intangible Cultural Heritage 14, **23–26**, 26–28, 86, 91
International Species Information System (ISIS) 40
international tourism destination 169
International Union for the Conservation of Nature 41
interspecies communication 143
Ireland-based NGO Animal Rights Action Network (ARAN) 79
Itsukushima Shrine **169**
IUCN Red List data 40

jallikattu
- animal cruelty 75
- animal ethics 74
- animal rights and animal welfare, concepts 76
- anti-ban campaign 82
- case study 78–79
- cultural relativism 77
- democratic formal justice 76
- development of research 76–77
- in heritage and cultural tourism 74
- Indian context 77–78
- non-consumptive wildlife tourism 75
- PETA 79
- posthumanism 74–75
- pursuing lawsuits 79–81
- in South Indian 78
- tourism policy 76

Japanese Heritage Tourism 141
- culture animals 143–144
- feline saviours 144
- tea ceremonies 52
- tourism research 168–169

kawaii
- animals and objects 141–142
- and heritage tourism 142

kawaiification 141, 188
Kishi station 144–147

'La Petatera' bullring 92
legislations 94
liveliness 159
lively capital concept 159–160
London Zoological Garden 35
loveliness 158
'Luwak Coffee Bali Plantation' 52

male bravery 78
man-bull contests 100
masculinity, hegemonic narratives 11
Mesoamerican Reef in Belize 26
methodology 171
Mexico
- animalist ideology 94–95
- Autonomous Communities 91
- bullfighter's identity 94
- bullfighting heritage ideology 91–94
- bullfights analysis as dissonant heritage 88–90
- cultural heritage 85
- cultural tourism 86
- heritage tourism 85–86
- historical, social and legal frames 90–91
- ideologies and discourses 91
- tourism discourse 89
- tourist industry 85
- traditionalist discourse 92
- transatlantic *tauromaquia* 90
- utilitarianism 89–90
- violence 91

Miko 144
mixed and cultural heritage sites 26–27
Miyajima deer 177–178
Miyajima island 171–181, **172**
- borderland politics 179–180
- as cultural tourism destination 171–172
- deer as active agency 180–181
- deer as wild 177–178
- discourse analysis 177
- habituated deer 178
- human-deer interaction
  - history 172–173
  - management 173–174
- location **172**
- maps **175**
- Miyajima Deer Protection Management Plan 177
- opposition groups 176–177
- residents 174–176
- spatial framing 177
- specialty of 173
- tourists 176
- transgression of boundaries 178
- wild concept 178–179

moral discrimination 102
moral justification of zoos 41
multiple stakeholders' views on deer 174–177
multi-species virtual ethnography 48
MUMAX 136
muted torment 1–5

National Autonomous University of Mexico 131
'Natsume' 149
natural capital and natural heritage 48
natural habitats 40
natural resources and biodiversity 163
natural space 179–180
natureculture framework 49–50
nekonomic kawaiification 149
nekonomics 140–141, 149
'neoliberal branded-self ' 51
neurosis 42
Nitama's cage at Kishi Station **147**
noise disturbance 14
non-consumptive wildlife tourism 75
nongjiales 163
non-human animal
- communities 132
- rights in tourism 75
- in tourism
  - critical approaches to heritage 10–12
  - ethical issues on the animals 12–13
  - from humanism to posthumanism 8–10
  - Intangible Cultural Heritage **23–26**, 26
  - mixed and cultural sites with animals **21–22**
  - natural sites with animals **15–20**
  - philosophical positions on moral consideration 9
  - tourism, animals and heritage 13–27
  - World Heritage Convention 26

noraneko 144
novillo 101

Oklahoma's tribal nations and indigenous connections with eagles 117–119
ontologization 26
Outstanding Universal Value 14

'Panda Homestay Village' 163–164
pandas in urban and rural China
- affective economies 157
- consumption gap in 162–165
- cute, affective economies of 158–159

pandas in urban and rural China (*Continued*)
 digital media economy 157
 emotional labour between bodies 157
 heritage of cute, consuming and building 161–162
 as lively capitals 159–160
 method 161
 panda-tourist relationship 160–161
 posthumanism studies 156–157
 sculpture 162
 self-domestication 157
panda-tourist relationship 160–161
patriarchalism 10
PETA (People for the Ethical Treatment of Animals) 75, 79
philanthropy 158
philosophical posthumanism 8
picador 103
political-ecological network 171
political ecology 170
politicization 74–75
pollution 14
post-anthropocentrism 103
posthumanism 4, 8–10, 58, 74–75, 103, 143, 171
posthumanist
 anti-speciesist perspective 69
 discourse 89
 zoos 35
posthumanist political ecology approach
 borderland politics 179–180
 boundaries, transgression of 178
 cultural tourism 168
 deer as active agency, evidence of 180–181
 ecofeminism 171
 human-deer interaction
  conflict 170–171
  history 172–173
  management *173*, 173–174
 humans and non-human animals 170
 human-wildlife conflict 169
 international tourism destination 169
 Itsukushima Shrine **169**
 Japanese tourism research 168–169
 methodology 171
 Miyajima deer 177–178
 Miyajima island 171–172, **172**
 multiple stakeholders' views on deer 174–177
 political ecology 170
 posthumanism 171
 red *torii* gate **169**
 society and environment 170
 wildlife 178–179
post-travel stage 13
Potawatomi cultural values and practices 121, 124–126
pre-Hispanic heritage 68
pre-travel stage 13
Prevention of Cruelty to Animals Act (PCA) 79
pro-*jallikattu* 79
'Project Tiger,' wildlife conservation programme 77–78
Pueblos Mágicos programme 129–131
puyas 103

recreation and tourism 35–38
Red List of Threatened Species of the IUCN 40
red *torii* gate **169**
reflective ethnographic approach 49–50
reinforce natural populations 40–41
'resistance to captivity' 55–56
resurgence
 as alternative 115–116
 paradigms 3
right to freedom 41–42

Sanada Maru 148
sanctuary 43
Sanjūrō 147–149
San Marcos National Fair (SMNF) 99–109
scientific research in zoos 38–39
scientific-social premises 2
ScrapeHero software 64
self-domestication 157, 160
selfie-culture 51
semantic renovation 86
semi-structured interviews 62
shikado (deer gates) 180–181
social capital 51
social-construction process 130
social contact 153
social-environmental interactions 170
solitary species 42
Sonora market, animal trade and dark tourism in
 animal rights groups 69
 animals in **65**
 as anti-place 68–70
 birds in crates **69**
 concepts found in visitors' reviews of **64**
 De la Cruz-Badiano Codex 68
 live birds sale in **66**
 live puppies sale in **67**
 location of **63**
 posthumanist anti-speciesist perspective 61
 pre-Hispanic heritage 68
 tourism
  and consumption of animals 63–68
  as consumption of otherness 62–63
 tourist imaginaries range 68
 traditional medicine and magic 61
Soto 143–144
South India, *jallikattu* in 78
spatial dichotomy 133
spatial framing 177
speciesism 4–5, 9, 14, 68, 102–103, 187
speculative accumulation 57
stereotypical behaviours 42

strays 143
stress 103
sustainable development concept 14
symbiotic existence 143

'Taiga Drama' 148
Tama Museum Kishi Station 145, **145**
Tama Train 145–146, **146**
Tamil culture 79
Tamil Nadu Regulation of Jallikattu Act 79
tangible cultural heritage 86
tastescape concept 48, 53
taurometre 93
techno-scientific knowledge 8
tenderness 4
tertiarization of economy 3
theming Chignahuapan 130–131
tienta 101
topophobia 68
toreros 101
tourism 1
  animals 5
  discourse 89
  geography and animal ethics 61–62
  and heritage 185
  policy, *jallikattu* 76
  -themed animal 4
tourist
  -animal relations 4, 28
  attraction, bullfighting 104
  industry 85
  interests in SMNF activities **106**
  -panda interactions 157
  -panda relationship 161
  trap 58
touristification 11, 14, 88
traditional Balinese landscape, desire to consume 50–51
traditionalist discourse 93
traditional mutualism 57
transatlantic *tauromaquia* 90
transcend barbarism 8
transgression of boundaries, Miyajima island 178
transhumanism 8
'trans-species relation' 159
Tribal Eagle Aviary Permit by Fish & Wildlife Service 120
Tribal-led Eagle Aviary Tours
  animal rights and species justice 113
  ethics and animal welfare 113
  human exceptionalism 113
  indigenous interpretations and indigenous-facilitated experiences of animals 114–115
  Indigenous resurgence and grounded normativity **125**
  Jennifer Randell, on tour and in conversation with 119–121
  methodology 116–117
  Oklahoma's tribal nations and indigenous connections with eagles 117–119
  posthumanist and Indigenous Potawatomi approaches *124*
  resurgence as alternative 115–116
turismo 1

'Ubud Sari Coffee Luwak Agrotourism' 52
Uchi 143–144
unbalanced power 188
United Nations Educational, Scientific and Cultural Organization (UNESCO)
  designations 14
  Intangible Cultural Heritage 86, 100
  World Cultural Heritage Sites 168–169, 188
  World Heritage List in 2006 156
Universal Declaration on Animal Rights 62
universalism 11
utilitarianism 12, 89–90

vessel trails 26
violence 91
virtual storytelling 51
volunteer conservation tourism 170

'web-scraping' of social-media platforms 61–62
welfarism 12
whaling 77
wild concept, Miyajima island 178–179
wilderness 158
wildlife 178–179
  preservation 38
  tourism 160
Wildlife (Protection) Act 77–78
World Association of Zoos and Aquariums (WAZA) 38
World Heritage Convention 14, 26–28
  mixed and cultural sites with animals **21–22**
  natural sites with animals **15–20**
World Heritage value 169
World Wildlife Fund (WWF) 4, 75, 156

xolotl 131

Yontama's cage at Idakiso Station **148**

zoocentrism 8
  philosophical posthumanism 26
zoonosis 69
zoos, legacy of
  animal sanctuaries 43–44

zoos, legacy of (*Continued*)
biobanks 39
captivity and animal rights 41–43
conservation 39–41
conservation programmes integration 43
as heritage tourist attractions 34
history 35
inbreeding in 41
moral justification of 41
posthumanist 35
recreation, tourism and education 35–38
scientific research 38–39
zootherapy 94

Printed and bound by CPI Group (UK) Ltd, Croydon, CR0 4YY

13/09/2026

14961113-0001